VOLUME ONE HUNDRED

Advances in
COMPUTERS
Energy Efficiency in Data Centers
and Clouds

VOLUME ONE HUNDRED

ADVANCES IN
COMPUTERS

Energy Efficiency in Data Centers and Clouds

Edited by

ALI R. HURSON

Missouri University of Science and Technology, Rolla, MO, USA

HAMID SARBAZI-AZAD

Sharif University of Technology, Tehran, Iran

ELSEVIER

AMSTERDAM • BOSTON • HEIDELBERG • LONDON
NEW YORK • OXFORD • PARIS • SAN DIEGO
SAN FRANCISCO • SINGAPORE • SYDNEY • TOKYO
Academic Press is an imprint of Elsevier

Academic Press is an imprint of Elsevier
50 Hampshire Street, 5th Floor, Cambridge, MA 02139, USA
525 B Street, Suite 1800, San Diego, CA 92101-4495, USA
The Boulevard, Langford Lane, Kidlington, Oxford OX5 1GB, UK
125 London Wall, London, EC2Y 5AS, UK

First edition 2016

Notices
Knowledge and best practice in this field are constantly changing. As new research and
experience broaden our understanding, changes in research methods, professional practices,
or medical treatment may become necessary.

Practitioners and researchers must always rely on their own experience and knowledge in
evaluating and using any information, methods, compounds, or experiments described
herein. In using such information or methods they should be mindful of their own safety and
the safety of others, including parties for whom they have a professional responsibility.

To the fullest extent of the law, neither the Publisher nor the authors, contributors, or editors,
assume any liability for any injury and/or damage to persons or property as a matter of
products liability, negligence or otherwise, or from any use or operation of any methods,
products, instructions, or ideas contained in the material herein.

ISBN: 978-0-12-804778-1
ISSN: 0065-2458

For information on all Academic Press publications
visit our web site at http://store.elsevier.com/

Working together
to grow libraries in
developing countries

www.elsevier.com • www.bookaid.org

CONTENTS

PREFACE

Traditionally, *Advances in Computers*, the oldest series to chronicle the rapid evolution of computing, annually publishes several volumes, each typically comprised of four to eight chapters, describing new developments in the theory and applications of computing. The theme of this 100th volume is inspired by the growth of data centers and their influence on our daily activities. The energy consumed by the large-scale warehouse computers and their environmental impacts are rapidly becoming a limiting factor for their growth. Many engineering improvements in the design and operation of data centers are becoming standard practice in today's state-of-the-art data centers, but many smart optimization opportunities still remain to be explored and applied. Within the domain of energy efficiency in data centers, this volume touches a variety of topics including sustainability, data analytics, and resource management. The volume is a collection of five chapters that were solicited from authorities in the field, each of whom brings to bear a unique perspective on the topic.

In Chapter 1, "Power Management in Data Centers: Cost, Sustainability, and Demand Response," Oo *et al.* articulate the existing system management techniques in data centers that mainly concern about power consumption and energy cost (either by cyber entities and/or physical entities such as cooling systems). The article then discusses novel approaches to manage systems in data centers based on metrics such as cost (online energy budgeting and thermal-aware scheduling), sustainability and environmental impact (exploitation of temporal diversity and optimization of water distribution), and demand response (real-time pricing and reserve auction).

In Chapter 2, "Energy-Efficient Big Data Analytics in Datacenters," Mehdipour *et al.* emphasize the effect of growth of the volume of data on data processing and capacity of the data centers articulating needs to develop new techniques for data acquisition, analyses, and organization with an eye to reduce the operational cost. Datacenter architecture, concept of big data, existing tools for big data analyses, and techniques for improving energy in datacenters are discussed. Finally, the concept of horizontal scaling as opposed to vertical scaling for handling large volume of processing while optimizing metrics such as power consumption, power supplies, and memory is addressed.

In Chapter 3, "Energy-Efficient and SLA-Based Resource Management in Cloud Data Centers," Sampaio and Barbosa present an overview about energy-efficient management of resources in cloud data centers, with quality of service constraints. Several techniques and solutions to improve the energy efficiency of computing systems are presented, and recent research for efficient handling of power and energy is discussed. Basic concepts pertaining to power, energy models, and monitoring the power and energy consumption are addressed. Sources of power consumption in datacenters are identified, and finally, a survey of resource management in datacenters is presented.

Chapter 4, "Achieving Energy Efficiency in Datacenters by Virtual Machine Sizing, Replication, and Placement," by Goudarzi and Pedram argues virtualization as a means to reduce energy consumption of datacenters. An overview of various approaches for consolidation, resource management, and power control in datacenters is presented. The chapter also includes a dynamic programming-based algorithm for creating multiple copies of a virtual machine without degrading performance and performing virtual machine consolidation for the purpose of datacenter energy minimization. It is shown that in comparison to the previous work, the aforementioned algorithm reduces the energy consumption by more than 20%.

Finally, in Chapter 5, "Communication-Awareness for Energy Efficiency in Datacenters," Nabavinejad, and Goudarzi concentrate on communication as a source of energy consumption in a datacenter. In a datacenter, this interest stems from the fact that about 10–20% of energy consumed by IT equipment is due to the network. An overview of network energy consumption in data centers is presented. Various techniques to reduce power consumption of network equipment are classified, reviewed, and discussed. A new approach is introduced, formulated, and simulated, and experimental results are presented and analyzed.

We hope that you find these articles of interest, and useful in your teaching, research, and other professional activities. We welcome feedback on the volume, and suggestions for future volumes.

ALI R. HURSON

Missouri University of Science and Technology, Rolla, MO, USA

HAMID SARBAZI-AZAD

Department of Computer Engineering,

Sharif University of Technology, Tehran, Iran and

School of Computer Science,

Institute for Research in Fundamental Sciences (IPM), Tehran, Iran

CHAPTER ONE

Power Management in Data Centers: Cost, Sustainability, and Demand Response

Thant Zin Oo*, Nguyen H. Tran*, Choong Seon Hong*, Shaolei Ren[†], Gang Quan[‡]

*Department of Computer Engineering, Kyung Hee University, Seoul, South Korea
[†]Department of Electrical and Computer Engineering, University of California, Riverside, Riverside, California, USA
[‡]Electrical and Computer Engineering Department, Florida International University, Miami, Florida, USA

Contents

Advances in Computers, Volume 100
ISSN 0065-2458
http://dx.doi.org/10.1016/bs.adcom.2015.10.001

Abstract

Due to demand for ubiquitous cloud computing services, the number and scale of data centers has been increasing exponentially leading to huge consumption of electricity and water. Moreover, data centers in demand response programs can make the power grid more stable and sustainable. We study the power management in data centers from perspectives of economic, sustainability, and efficiency. From economic perspective, we focus on cost minimization or budgeting of data centers. From sustainability point of view, we look at water and carbon footprint in addition to energy consumption. Finally, we study demand response between data centers and utilities to manage the power grid efficiently.

NOMENCLATURE

CFD computational fluid dynamics
CPU central processing unit
DC(s) data center(s)
DR demand response
DSP(s) data center service provider(s)
DWB deciding winning bids
eBud energy Budgeting
EWIF energy water intensity factor
GLB-Cost Geographical Load Balancing for Cost minimization
GLB-WS Geographical Load Balancing for Water Sustainability
GreFar Geographical Cost Minimization with Thermal Awareness
iCODE incentivizing COlocation tenants for DEmand response
ISP(s) Internet service provider(s)
PAR peak-to-average ratio
PerfectPH Perfect Prediction Heuristic
PUE power usage effectiveness

PV photovoltaic
QoS quality of service
VM(s) virtual machine(s)
WACE minimization of WAter, Carbon and Electricity cost
WUE water usage effectiveness

1. INTRODUCTION

Demand for ubiquitous Internet and cloud services has led to construction of gigantic data centers (DCs), which contain both cyber assets (e.g., servers, networking equipment) and physical assets (e.g., cooling systems, energy storage devices). With dramatically surging demand of cloud computing services, service providers have been expanding not only the scale but also the number of DCs that are geographically distributed.

With this increasing trend, DCs have become large-scale consumers of electricity. A study shows that DCs account for 1.7–2.2% of the total electricity usage in the United States as of 2010 [1]. Another study shows that many DC operators paid more than $10 million [2] on their annual electricity bills, which continues to rise with the flourishing of cloud computing services. Moreover, companies like Google and Microsoft spend a large portion of their overall operational costs on electricity bills [3]. This energy consumption is often labeled as "brown energy" due to its carbon-intensive sources. Decreasing the soaring energy cost is imperative in large DCs. Some works have shown that DC operators can save more than 5–45% [4] operation cost by leveraging time and location diversities of electricity prices.

This demand for electricity also has profound impacts on the existing ecosystem and sustainability. A less-known fact about DCs is that they are extremely "thirsty" (for cooling), consuming millions of gallons of water each day and raising serious concerns amid extended droughts. In light of these, tremendous efforts have been dedicated to decreasing the energy consumption as well as carbon footprints of DCs (see Refs. [5–7] and references therein). Consequently, DC operators are nowadays constantly urged to cap the increasing energy consumption, either mandated by governments in the form of Kyoto-style protocols or required by utility companies [5, 7–10]. These growing concerns have made energy efficiency a pressing issue for DC operation.

1.1 Existing System Management Techniques in Data Centers

A DC is essentially a server farm that consists of many servers. These servers may have different capacity limits (constraints). Generally, the system management context of DCs used to involve only the power (energy) management. The power (energy) inside a DC is consumed by either the cyber assets (IT equipment) or the physical assets (cooling system). In cyber assets, each server is a block of resources such as processing power (CPU), memory, storage (hard disk), and network bandwidth. To deal with the physical assets of the DC, power usage effectiveness (PUE), the ratio of total amount of energy consumed by a DC to the energy consumed by its IT equipment, is used as a metrics measuring the energy (e.g., cooling) efficiency of the DC. An ideal PUE is 1.0 which means all the energy is consumed by cyber assets.

Power management in a DC can be described as an optimization problem with cyber–physical constraints. For a DC, the constraints are similar to those of server farm from legacy client–server system. With the implementation of virtualization, the process is more streamlined and efficient. Nonetheless, from the DC operators' perspective, the optimization goals (objectives) from the legacy system remain relevant in a DC. Three main objectives of power management issues in a DC are reducing operating (electricity) cost [2, 4, 11, 12], reducing response time [13], and reducing energy consumption [5, 7, 14].

From the cost reduction perspective, "power proportionality" via dynamically turning on/off servers based on the workloads has been studied extensively [4, 6, 12, 15]. In these works, the authors primarily focus on balancing between energy cost of DC and performance loss through dynamically provisioning server capacity. In Ref. [2], the authors proposed geographical load balancing among multiple distributed DCs to minimize energy cost. In Ref. [5], the authors cap the long-term energy consumption based on predicted workloads. In Refs. [7, 14], the authors propose to reduce brown energy usage by scheduling workloads to DCs with more green energies. Cyber–physical approaches to optimizing DC cooling system and server management are also investigated [16, 17]. To further reduce the electricity cost, the advantage of geographical load balancing is combined with the dynamic capacity provisioning approach [4].

1.2 Novel System Management Techniques in Data Centers

The legacy objectives, such as minimizing cost, energy consumption, and delay, remain relevant and thus are inherited for current and future DC system management techniques. However, in some of our recent works, the system management context is expanded to include other physical assets of the DCs, i.e., temperature-aware scheduling [18] and water usage minimization [19, 20]. With the advent of cloud computing, there is a growing trend toward system management techniques for the geographically distributed DCs [6, 7]. Some of our recent works consider energy efficiency for a geographically distributed set of DCs [18, 20–22], instead of a single DC. We next characterize some of our recent works.

In terms of DC cost minimization, we will present two following ideas:

- **Online energy Budgeting** (eBud) [23]: The objective is to minimize DC operational cost while satisfying a long-term energy cap. Formulated as an optimization problem with a fixed budget constraint, we employed Lyapunov optimization technique to solve it. Then, we proposed an online algorithm called "eBud" which uses a virtual queue as a control mechanism.

- **Thermal-aware scheduling for geographically distributed DCs** (GreFar) [18]: In addition to normal cyber constraints, we added sever inlet temperature constraint which prevents server overheating. The objective is to minimize operational cost of geographically distributed DCs. The problem is formulated as scheduling batch jobs to multiple geographically distributed DCs. We also employed Lyapunov optimization technique and proposed an online scheduling algorithm called "GreFar."

In terms of sustainability, we will present two following ideas:

- **Sustainability: exploiting temporal diversity of water efficiency** (WACE) [19]: Not much attention is paid to DC water consumption (millions of gallons of water each day) which raise serious concern for sustainability. We adopted the characteristic of DC-time-varying water efficiency into server provisioning and workload management. The objective is to minimize operational cost of the DC. We formulated the problem into minimizing the total cost under resource (water, carbon, and energy) and quality of service (QoS) constraints. As in our other works [18, 23], we apply Lyapunov optimization technique. Then, we proposed an online algorithm called "WACE" (minimization of WAter, Carbon and Electricity cost), which employs a virtual queue. WACE dynamically adjusts server provisioning to reduce the water consumption by deferring delay-tolerant batch jobs to water-efficient time periods.

- **Sustainability: optimizing water efficiency in distributed DCs** (GLB-WS) [20]: We identify that water efficiency of a DC varies significantly not only over time (temporal diversity) but also with location (spatial diversity). The objective is to maximize the water efficiency of distributed DCs. We formulated the problem into a maximizing the water efficiency under resource, QoS, and budget constraints. Afterward, the problem was transformed into a linear-fractional programming problem [24]. Then, we provide an iterative algorithm, called "GLB-WS" (Geographical Load Balancing for Water Sustainability), based on bisection method. GLB-WS dynamically schedules workloads to water-efficient DCs for improving the overall water usage effectiveness (WUE) while satisfying the electricity cost constraint.

In terms of demand response (DR), we will present two following ideas:

- **DR of geo-distributed DCs: real-time pricing game approach** [22]: DR program, a feature of smart grid, reduces a large electricity demand upon utility's request to reduce the fluctuations of electricity demand (i.e., peak-to-average ratio (PAR)). Due to their huge energy consumption, DCs are promising participants in DR programs, making power grid more stable and sustainable. In this work, we modeled the DR between utilities and geographically distributed DCs using a dynamic pricing scheme. The pricing scheme is constructed based on a formulated two-stage Stackelberg game. Each utility connected to the smart grid sets a real-time price to maximize its own profit in Stage I. Based on these prices, the DCs' service provider minimizes its cost via workload shifting and dynamic server allocation in Stage II. The objective is to set the "right prices" for the "right demand."

- **DR of colocation DCs: a reverse auction approach** (iCODE) [21]: In this work, we focus on enabling colocation DR. A colocation DC hosts servers of multiple tenants in one shared facility. Thus, it is different from a owner-operated DCs and suffers from "split incentive." The colocation operator desires DR for financial incentives but has no control over tenants' servers, whereas tenants who own the servers may not desire DR due to lack of incentives. To break "split incentive," we proposed an incentive mechanism, called "iCODE" (incentivizing COlocation tenants for DEmand response), based on reverse auction. Tenants can submit energy reduction bids to colocation operator and will be financially rewarded if their bids are accepted. We build a model to represent how each tenant decides its bids and how colocation operator decides winning bids. The objective is to reduce colocation energy consumption in colocation DCs. iCODE employs branch and bound technique to yield a suboptimal solution with a reasonably low complexity [25].

2. COST MINIMIZATION IN DATA CENTERS

In this section, we will discuss about minimizing DC operational costs by managing its energy consumption. This section is divided into two parts. The first part, Section 2.2, deals with optimization of a single DC, whereas the second part, Section 2.3, exploits the spatial diversity of distributed DCs.

2.1 Related Work on DC Optimization and VM Resource Management

Research in DC operation optimization is recently receiving more attention. From economic perspective, data center service providers (DSPs) want to reduce operation cost, for example, cutting electricity bills [4, 7, 12]. On the other hand, DSPs may also want to reduce latency in terms of performance [6, 13]. "Power proportionality" via dynamically switching on/off servers based on workloads is a well-researched approach to reduce energy cost of DSs [12]. With the rise of virtualization in DCs, researches have also focus on virtual machine (VM) resource management. In Refs. [26, 27], the authors studied the optimum VM resource management in the cloud. Reference [28] studied admission control and dynamic CPU resource allocation to minimize the cost with constraints on the queuing delay for batch jobs. In Ref. [29], authors explored autonomic resource allocation to minimize energy in a multitier virtualized environment. References [30–32] studied various dynamic VM placement and migration algorithms. Some of the mentioned works can be combined with our proposed solutions. However, these studies assume that server CPU speed can be continuously chosen, which is not practically realizable because of hardware constraints.

2.2 Online Energy Budgeting

DCs are large consumers of electricity and thus have environmental impacts (e.g., carbon emission). Therefore, energy consumption has become an important criteria for DC operation. A pragmatic move for IT companies is to cap the long-term energy consumption [5]. In Ref. [23], we study long-term energy capping for DCs with a proposed online algorithm called eBud (energy Budgeting) that have many functionalities such as number of active servers decisions, VM resource allocation, and workload distribution. Existing researches either focus only on VMs placement and/or migration [30–32], or just eBud without virtualized systems [33].

2.2.1 Related Work on Power Budgeting and Energy Capping

It is crucial for the servers to optimally allocate the limited power budget due to the expensive cost associated with DC peak power [13]. "Power budgeting" optimally allocates the limited power budget to (homogeneous) servers to minimize the latency based on a queuing theoretic model [13, 34]. "Energy Budgeting" is related to but different from well-studied "power budgeting." eBud remains a relatively less explored area because of the challenge in making decision under lack of information on future workloads. Furthermore, eBud requires an online algorithm to deal with the uncertainty of future workloads. In Refs. [5, 10, 35], the authors relied on long-term prediction of the future information to cap the monthly energy cost. In comparison, eBud guarantees the average cost with bounds on the deviation from energy capping constraint. The simulation results [23] also demonstrate the benefits of eBud over the existing methods empirically. In Ref. [33], authors studied eBud for a DC, but they focus on dynamically switching on/off servers and hence do not apply to virtualized DCs which require a set of holistic resource management decisions. Moreover, eBud improves [23, 33] by introducing a self-configuration approach to dynamically update the cost capping parameter.

2.2.2 Problem Formulation

We focus on how to minimize the long-term virtualized DC's operational cost subject to an energy capping as follows:

$$\textbf{P1}: \quad \text{minimize}: \quad \frac{1}{T}\sum_{t=0}^{T-1} g\left(\mathbf{m}(t), \mathbf{x}(t), \boldsymbol{\lambda}(t)\right), \tag{1}$$

$$\text{subject to}: \quad \sum_{t=0}^{T-1} p\left(\mathbf{u}(t), \mathbf{m}(t)\right) \leq Z. \tag{2}$$

In this problem, at time t of a total considered T time slot, let $p(\mathbf{u}(t), \mathbf{m}(t))$ denotes the total power consumption of the DC with a vector of average server utilization $\mathbf{u}(t)$, and a vector of number of turned-on servers $\mathbf{m}(t)$, of total I physical server's types. The total electricity cost can be expressed as $e(\mathbf{u}(t), \mathbf{m}(t)) = w(t) \cdot p(\mathbf{u}(t))$ where $w(t)$ denotes the electricity price. Furthermore, let $\lambda_{i,j}$ denote a type j of a total different J types of workloads processed at type-i server and $\mu_{i,j}$ denotes the service rate of $\text{VM}_{i,j}$ at type-i server serving type-j workloads. Then, the total delay cost of all

$VM_{i,j}$, $\forall i,j$, at time t is represent as $d(\lambda(t),\mu(t),\mathbf{m}(t))$, where $\lambda(t)$ and $\mu(t)$ are vectors of $\lambda_{i,j}$ and $\mu_{i,j}$, $\forall i,j$, respectively. Hence, the time–average objective of problem **P1** is constructed from the total DC's cost at time t which is presented as follows:

$$g\left(\mathbf{m}(t),\, \mathbf{x}(t),\, \lambda(t)\right) = e(\mathbf{u}(t),\mathbf{m}(t)) + \beta \cdot d\left(\lambda(t),\, \mu(t),\mathbf{m}(t)\right), \qquad (3)$$

where $\mathbf{x}(t)$ represents for VM resource allocation, and $\beta \geq 0$ is weighting parameter for delay cost relative to the electricity cost [6, 7]. Furthermore, the constraint of **P1** indicates an energy budget Z.

2.2.3 Scheduling Algorithm (eBud)

ALGORITHM 1 eBud

1: Input $a(t)$ and $w(t)$ at the beginning of each time $t=0$, $1,...,$ $T-1$
2: Dynamically updating V if necessary
3: Choose $\mathbf{m}(t)$, $\mathbf{x}(t)$, and $\lambda(t)$ subject to constraints described in Ref. [23] to solve

$$\mathbf{P2}: \text{minimize}: \ V \cdot g\left(\mathbf{m}(t),\mathbf{x}(t), \lambda(t)\right) + q(t) \cdot p\left(\mathbf{u}(t),\mathbf{m}(t)\right) \qquad (4)$$

4: At the end of time slot, update $q(t)$ according to (5).

To solve **P1**, an online algorithm, called eBud, was developed, based on Lyapunov optimization [36]. eBud is presented in Algorithm 1. Basically, this technique introduces a virtual budget deficit queue that monitors the deviation from the long-term target, and gradually nullifies the energy capping deviation:

$$q(t+1) = \left[q(t) + p\left(\mathbf{u}(t),\mathbf{m}(t)\right) - \frac{Z}{T}\right]^{+}. \qquad (5)$$

The proof on the performance of eBud is available in Ref. [37]. The shortcoming of eBud is its slow convergence rate.

As described in Ref. [37, Theorem 1], V places the limit on the maximum deviation from the long-term energy capping target. Furthermore, V decides how close the average cost achieved by eBud will be to the lower bound.

2.2.4 Performance Evaluation of eBud

We present trace–based simulation of a DC to validate their analysis and performance of eBud, where details of simulation settings and data sets are in

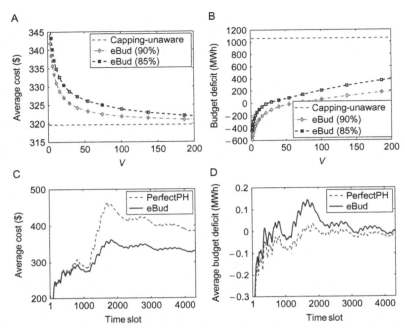

Figure 1 Impact of V to eBud and comparison of eBud with prediction-based methods. (A) Average cost approaches lower bound as V increases, (B) zero budget deficit is achieved by eBud at different V for different budget, (C) comparison of average cost, and (D) comparison of average budget deficit.

Ref. [23]. Figure 1A and B displays the impact of V. With the increasing V, the average cost decreases, while the budget deficit increases. For a large value of V, average cost minimization has a greater weight than energy capping. Figure 1C and D displays the comparison between eBud and the prediction-based Perfect Prediction Heuristic (PerfectPH) [23] in terms of the average cost and average budget deficit per time slot. The operational costs of PerfectPH are higher than eBud because of short-term prediction while allocating the energy budget.

2.3 Geographical Cost Minimization with Thermal Awareness (GreFar)

Energy efficiency and electricity prices vary temporally and spatially, and thermal management in DCs is important due to its excessive amount of heat [38]. Thus, how to exploit the electricity price variations across time and locations with server temperature consideration is studied in Ref. [18].

2.3.1 Related Work on Thermal-Aware Resource Management

Thermal-aware resource management can be classified as temperature-reactive and temperature-proactive approaches. In temperature-reactive approaches, the algorithm reacts to avoid server overheating based on the observed real-time temperature [39–41]. Reference [39] dynamically schedules workloads to cool servers based on the instantaneously observed temperature. Reference [40] optimally allocates power in a DC for MapReduce workloads taking into consideration the impact of temperature on server performance. In Ref. [41], authors presented a cyber–physical approach monitoring the real-time temperature distribution in DCs, enabling temperature-reactive decisions. In temperature-proactive approaches, the algorithm explicitly considers the impact of resource management decisions on the temperature increase [42–45]. Direct introduction of computational fluid dynamics (CFD) models for temperature prediction incurs a prohibitive complexity [42]. Thus, a less complex yet sufficiently accurate heat transfer model was developed in Ref. [43]. In Ref. [46], the authors disregard latency to maximize the total capacity of a DC subject to a temperature constraint. In Refs. [44, 45], the authors presented software/hardware architectures for various types of thermal management in high-performance computing environments. In comparison, GreFar does not require any prior knowledge or assume any (stationary) distributions of the system dynamics. GreFar is a provably efficient solution that minimizes the energy-fairness cost with constraints on temperature and queuing delay.

2.3.2 Problem Formulation

We study the long-term cost minimization problem as follows:

$$\textbf{P3}: \quad \underset{\mathbf{z}(t)}{\text{minimize:}} \quad \frac{1}{T}\sum_{\tau=0}^{T-1} g\left(\tau\right), \tag{6}$$

$$\text{subject to:} \quad a_j(t) = \sum_{i\in\mathcal{D}_j} r_{i,j}(t), \quad \forall j \in \mathcal{J} \tag{7}$$

$$\max \mathbf{L}_i^{in}(t) = \mathbf{L}_i^{sup}(t) + \max\left[\mathbf{D}_i(t)\cdot\mathbf{b}_i(t)\right] \leq L_i^{max}, \tag{8}$$

$$\sum_{j=1}^{J} h_{i,j}(t)d_j \leq \sum_{k=1}^{K_i(t)} u_{i,k}(t)s_k \leq \sum_{k=1}^{K_i(t)} s_k. \tag{9}$$

In this problem, the time-average objective is constructed from the instantaneous energy-fairness cost function defined as follows:

$$g(t) = e(t) - \beta \cdot f(t) = \sum_{i=1}^{N} e_i(t) - \beta \cdot f(t), \qquad (10)$$

where $e_i(t)$ is the energy cost for cooling systems and processing the scheduled batch jobs in DC i, $f(t)$ is the fairness function of the resource allocation, and $\beta \geq 0$ is a energy-fairness transfer parameter.

Furthermore, the server inlet temperature constraint is in (8), and the maximum amount of work that can be processed in DC i is represented in constraint (9) with control decisions $\mathbf{z}(t) = \{r_{i,j}(t), h_{i,j}(t), u_{i,k}(t)\}$ at time t represent for the number of type-j jobs scheduled to DC i: $r_{i,j}(t)$, the number of type-j jobs processed in DC i: $h_{i,j}(t)$, and the utilization of server k in DC i: $u_{i,k}(t)$. More details of parameters and functions are provided in Ref. [18].

2.3.3 Online Algorithm (GreFar)

An online algorithm, called "GreFar," is proposed to solve the optimization problem **P3**. GreFar, described in Algorithm 2, intuitively trades the delay for energy-fairness cost saving by using the queue length as a guidance for making scheduling decisions: jobs are processed only when the queue length becomes sufficiently large and/or electricity prices are sufficiently low. The queue dynamics is expressed as

$$q_{i,j}(t+1) = \max\left[q_{i,j}(t) - h_{i,j}(t), 0\right] + r_{i,j}(t), \qquad (11)$$

where $q_{i,j}(t)$ is the queue length for type-j jobs in DC i during time t.

ALGORITHM 2 GreFar Scheduling Algorithm

1: At the beginning of every time slot t, observe the DC state and the vector of current queue states;

2: Choose the control action subject to (7)–(9) to solve

$$\mathbf{P4} : \text{minimize} : V \cdot g(t) + \sum_{j=1}^{J} \sum_{i \in \mathcal{D}_j} q_{i,j}(t) \cdot \left[r_{i,j}(t) - h_{i,j}(t)\right], \qquad (12)$$

where the cost $g(t)$ is defined in (10).

3: At the end of time slot, update $q_{i,j}(t)$ according to (11).

2.3.4 Performance Evaluation of GreFar

GreFar is compared with the algorithm defined in Ref. [47] which is referred to as T-unaware. Full details on the simulation setup are described in Ref. [18]. Figure 2 depicts the total cost at each time slot for GreFar and

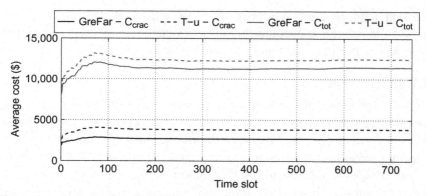

Figure 2 Average total and cooling costs.

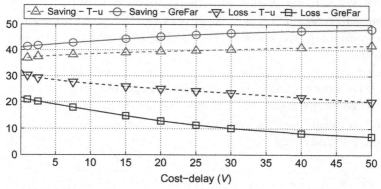

Figure 3 Saving and loss functions for different values of V.

T-unaware. Primarily, the savings are in the reduction of the cooling cost, as shown in Fig. 2, where only cooling costs for both algorithms are displayed. Thus, GreFar can reduce the overall cost acting on the cooling system. The *Saving* value represents cost reduction with respect to the most expensive case, whereas the *Loss* value represents the cost inefficiency with respect to the least expensive case. *Saving* and *Loss* values are depicted in Fig. 3 as a function of V. The results highlight that GreFar performs better than T-unaware.

3. SUSTAINABILITY: WATER EFFICIENCY

In Section 2, we discussed about operational cost minimization of DCs in terms of energy consumption and its importance. In this section, we will explore the importance of water efficiency and sustainability. In Sections 4 and 5, we will further present system management techniques to reduce water usage and increase water efficiency in DCs.

3.1 Motivation

DCs are well known for consuming a huge amount of electricity accounting for carbon emissions in electricity generation. However, a less known fact is that DCs' physical assets evaporate millions of gallons of water each day for rejecting server heat. For instance, water consumption of cooling towers in AT&T's large DC facilities in 2012 is 1 billion gallons, approximately 30% of the entire company's water consumption [48]. Furthermore, DCs also indirectly consume a huge quantity of water embedded in generating electricity [49, 50]. For example, in the United States, for every 1 kWh electricity generation, an average of 1.8 L of water is evaporated, even disregarding the water-consuming hydropower [49–51]. Figure 4 depicts a typical water-cooled DC.

A survey done by Uptime Institute [52] shows that a significant portion of large DCs use cooling towers (over 40%), despite the existence of various types of cooling systems. For example, in low-temperature regions, cold outside air can be used for cooling DCs. Moreover, to obtain green certifications [52] and tax credits [53], water conservation is essential. In addition,

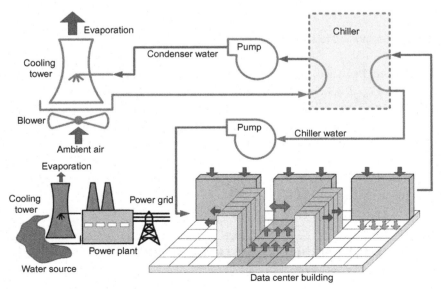

Figure 4 Water consumption in data center.

water is also important in generating electricity (e.g., thermoelectricity, nuclear power) [50, 51].

In spite of the relationship between water and energy [50, 51], the existing management techniques for minimizing DC energy consumption cannot minimize water footprint. This is because the physical characteristics of DC cooling systems and electricity generation are neglected. In order to reduce water footprint, when to consume energy must be considered in addition to minimizing energy consumption.

3.2 Related Work

In Ref. [54], the authors point out the critical issue of water conservation. In Refs. [55, 56], the authors develop a dashboard to visualize the water efficiency. No effective solutions have been proposed toward water sustainability in DCs. In most of the current works, improved "engineering" solutions have to be considered such as installing advanced cooling system (e.g., outside air economizer), by using recycled water, and by powering DCs with on-site renewable energies to reduce electricity consumption [57, 58]

These engineering-based approaches suffer from several limitations. First, they require appropriate climate conditions and/or desirable locations that are not applicable for all DCs (e.g., "free air cooling" is suitable in cold areas such as Dublin where Google has one DC [58]). Second, they do not address indirect off-site water consumption (e.g., on-site facilities for treating industry water or seawater save freshwater but may consume more electricity [56]). Third, some of these techniques often require substantial capital investments that may not be affordable for all DC operators.

Preliminary work on water sustainability via algorithm design is presented in Ref. [59], where the authors minimize water footprint via resource management. In comparison, WACE and GLB-WS focus primarily on minimizing water usage via algorithm design. WACE was designed to holistically minimize electricity cost, carbon emission, and water footprint by leveraging the delay tolerance of batch jobs and temporal diversity of water efficiency. Proposed GLB-WS explicitly focuses on water efficiency that is becoming a critical concern in future DCs. WACE and GLB-WS may not possibly outperform all the existing GLB techniques in every aspect (e.g., GLB-WS incurs a higher electricity cost than GLB-Cost that particularly minimizes the cost). In light of the global water shortage trend, incorporating water efficiency is increasingly essential in future research efforts.

3.3 Challenge

3.3.1 Exploiting Temporal Diversity of Water Efficiency (WACE) [19]

Water footprint is included in the problem formulation as an integral metric for DC operation. Intuitively, by deferring batch workloads to time periods with better water efficiency, the temporal diversity of water is exploited. During time periods with low water efficiency, only the necessary servers to serve interactive workloads will be kept online and others shutdown. The challenge is the difficulty in determining which time periods are water-efficient without future information due to the time-varying nature of water efficiency, job arrivals, carbon emission rate, and electricity price. To address the challenge, an online algorithm called "WACE" (minimization of WAter, Carbon and Electricity cost) was proposed. The objective is to reduce water footprint, while also decreasing the electricity cost and carbon footprint.

3.3.2 Optimizing Water Efficiency in Distributed DCs (GLB-WS) [20]

Water efficiency varies significantly over time and also by location. Intuitively, the total water footprint can be reduced by deciding "when" and "where" to process workloads. The challenge is to include the *temporal* and *spatial* diversities of DC water efficiency into the problem formulation. To address the challenge, a novel online algorithm called "GLB-WS" (Geographical Load Balancing for Water Sustainability) was proposed. The objective is to improve overall water efficiency by scheduling workloads to water-efficient DCs while maintaining the electricity cost constraint.

3.4 Water Consumption in Data Centers

Water is consumed both directly (i.e., by cooling system) and indirectly (i.e., by electricity generation) in DCs. DCs' direct water usage is mainly for the cooling systems (especially water-cooled chiller systems employ evaporation as the heat rejection mechanism). For example, even with outside air cooling, water efficiency of Facebook's DC in Prineville, OR is still 0.22 L/kWh [57], whereas eBay uses 1.62 L/kWh (as of May 29, 2013) [60]. The process of thermoelectricity generation also employs evaporation for cooling and hence is a large consumer of water [49, 57]. This water usage accounts for DCs' indirect water usage. While some types of electricity generation use no water (e.g., by solar photovoltaics (PVs) and wind), "water-free" electricity accounts for a very small portion in the total electricity generation capacity (e.g., less than 10% in the United States [49]). Indirect water efficiency is quantified in terms of "EWIF" (energy water intensity factor), which measures the amount of water consumption per kWh electricity (e.g., the U.S. national average EWIF is 1.8 L/kWh [49, 50]).

To evaluate the water efficiency of a DC, Green Grid has developed a metric, called water usage effectiveness, which is defined as [50],

$$WUE = \frac{\text{direct water usage} + \text{indirect water usage}}{\text{IT equipment energy}}. \tag{13}$$

The smaller the value of WUE, the more water efficient a DC is and the theoretical minimum WUE is zero (L/kWh).

DC water efficiency displays temporal and spatial diversities. Figure 5 depicts the average daily WUE over a 90-day period where the drastic changes of WUE can be seen over time. Depending on the methods of generation, EWIFs differ depending on the sources (e.g., thermal, nuclear, hydro) [62] as in Fig. 6C. Excluding hydroelectric, nuclear electricity consumes the most water, followed by coal-based thermal electricity and then solar PV/wind electricity [49]. The average EWIF exhibits a temporal diversity due to varying peak/nonpeak demand [14]. Figure 6A depicts the time-varying EWIF for California, calculated based on Ref. [62], and California energy fuel mixes [3]. Both direct and indirect WUEs vary substantially across different geographic locations. Figure 6D depicts the spatial diversity of the average EWIF in state level. By comparing the direct WUEs of Facebook's DCs in Prineville, OR and Forest City, NC, a significant variation between the two locations can be observed. A similar case of spatial diversity in direct WUE can also be seen by comparing the direct WUEs of Facebook's and eBay's DCs [57, 60].

Figure 5 Direct WUE of Facebook's data center in Prineville, OR (February 27 to May 28, 2013) [57].

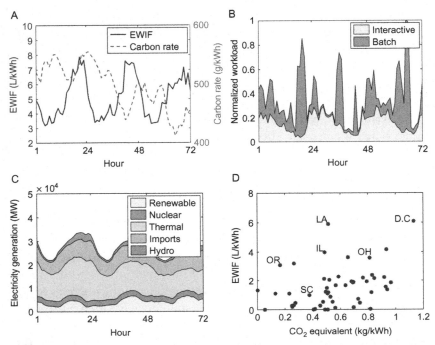

Figure 6 Water–carbon efficiency, workload trace, CA electricity fuel mix, and EWIF versus carbon emissions. (A) EWIF and carbon emission rate in California; (B) workload trace [15]; (C) CA electricity fuel mix on June 16 and 17, 2013 [3]; and (D) state-level EWIF versus CO_2 emissions in United States [50, 61].

4. SUSTAINABILITY: EXPLOITING TEMPORAL DIVERSITY OF WATER EFFICIENCY (WACE)

In Section 3, we discussed about the importance of water and sustainability. In this section, we will present a system management technique which exploits the temporal diversity of water and carbon [19]. We incorporated water and carbon footprints as constraints in the optimization problem formulation.

4.1 System Model of WACE

Let T be the total number of time slots in the period of interest. There are typically two types of workloads in DCs: delay-tolerant batch workloads (e.g., back-end processing, scientific applications) and delay-sensitive interactive workloads (e.g., web services or business transaction applications). Let $a(t) = [0, a_{max}]$ denote the amount of batch workload arrivals at time t [63, 64].

Let $r(t)$ denote amount of on-site renewable energy a DC has, e.g., by solar panels [65]. Let there be a total of $M(t)$ homogeneous servers that are available for processing batch jobs at time t. Servers can operate at different processing speeds and incur different power [66]. Let s represents the speed chosen for processing batch workloads from an array of finite processing speeds denoted by $S = \{s_1, ..., s_N\}$. The average power consumption of a server at time t can be expressed as $\alpha \cdot s(t)^n + p_0$ [66, 67], where α is a positive factor and relates the processing speed to the power consumption, n is empirically determined (e.g., between 1 and 3), and p_0 represents the power consumption in idle or static state. Server energy consumption by interactive workloads can be modeled as an exogenously determined value $p_{int}(t)$. The energy consumption by batch workloads can be written as $p_{bat}(t) = m(t) \cdot [\alpha \cdot s(t)^n + p_0]$. Hence, the total server energy consumption can be formulated as

$$p(t) = p_{bat}(t) + p_{int}(t). \tag{14}$$

Thus, given on-site renewable energy $r(t)$, the DC's electricity usage at time t is $[\gamma(t)p(t) - r(t)]^+$, where $[\cdot]^+ = \max\{\cdot, 0\}$ and $\gamma(t)$ is the PUE factor capturing the non-IT energy consumption. Let $u(t)$ denote the electricity price at time t, and hence, the electricity cost is $e(t) = u(t) \cdot [\gamma(t)p(t)r(t)]^+$, where $[\gamma(t)p(t)r(t)]^+$ is the DC electricity usage. The average EWIF can be calculated as [19]:

$$\varepsilon_I(t) = \frac{\sum_k b_k(t) \times \varepsilon_k}{\sum_k b_k(t)} \tag{15}$$

where $b_k(t)$ denotes the amount of electricity generated from fuel type k, and ε_k is the EWIF for fuel type k. The total water consumption at time t is given by

$$w(t) = \varepsilon_D(t) \cdot p(t) + \varepsilon_I(t) \cdot [\gamma(t) \cdot p(t) - r(t)]^+ \tag{16}$$

where $\varepsilon_D(t)$ is direct WUE at time t, $p(t)$ is the server power, $\gamma(t)$ is PUE, and $r(t)$ is available on-site renewable energy. The average carbon emission rate can be calculated as in Ref. [14]. Figure 6A depicts the time-varying carbon emission rate for California in which carbon emission efficiency does not align with EWIF (i.e., indirect water efficiency). The total carbon footprint of the DC at time t can be expressed as $c(t) = \phi(t) \cdot [\gamma \cdot p(t) r(t)]^+$.

4.2 Problem Formulation of WACE

The goal is to minimize the electricity cost while incorporating costs of carbon emission and water consumption [19]. Thus, a parameterized total cost function is constructed as follows:

$$g(t) = e(t) + h_w \cdot w(t) + h_c \cdot c(t), \tag{17}$$

where $h_w \geq 0$ and $h_c \geq 0$ are weighting parameters for water consumption and carbon emission relative to the electricity cost. The objective of WACE [19] is to minimize the long-term average cost expressed as $\bar{g} = \sum_{t=0}^{T} g(t)$, where T is the total number of time slots in the period of interest. The problem formulation for batch job scheduling is presented as follows [19]:

$$\textbf{P5}: \quad \underset{s(t), m(t)}{\text{minimize:}} \quad \bar{g} = \frac{1}{T} \sum_{t=0}^{T-1} g(s(t), m(t)), \tag{18}$$

$$\text{subject to :}$$
$$\text{\# of online servers :} \quad 0 \leq m(t) \leq M(t), \tag{19}$$
$$\text{Supported server speeds :} \quad s(t) \in \mathcal{P} = \{s_0, s_1, \ldots, s_N\}. \tag{20}$$
$$\text{Batch jobs not dropped :} \quad \bar{a} < \bar{b}, \tag{21}$$
$$\text{Data center capacity :} \quad b(t) = m(t) \cdot s(t), \tag{22}$$

where $\bar{a} = \sum_{t=0}^{T-1} a(t)$ and $\bar{b} = \sum_{t=0}^{T-1} b(t)$ are the long-term average workload arrival and allocated server capacity, respectively. The constraint (22) states the relation between processed batch jobs and server provisioning.

4.3 Online Algorithm (WACE)

To enable an online algorithm, the constraint (21) is removed. A batch job queue is maintained to store unfinished jobs [19]. Assuming that $q(0) = 0$, the job queue dynamics can be written as

$$q(t+1) = [q(t) - b(t)]^+ + a(t), \tag{23}$$

where $[\cdot]^+ = \max\{\cdot, 0\}$, $a(t)$ quantifies batch job arrivals, and $b(t)$ indicates the amount of processed jobs.

By intuition, when the queue length increases, the DC should increase the number of servers and/or server speed to reduce the queue backlog to reduce long delays. Therefore, the queue length is integrated into the objective function, as described in Algorithm 3. In (24), how much emphasis the optimization gives on the resource provisioning $b(t)$ for processing batch jobs is determined by the queue length. WACE can be implemented as a purely online algorithm which only requires the currently available information. The parameter $V \geq 0$ in line 2 of Algorithm 3, referred to cost–delay parameter, functions as a trade-off control knob. The larger the value of V, the smaller impact of the queue length on optimization decisions.

WACE is simple but provably efficient, even compared to the optimal offline algorithm that has future information. Based on the recently developed Lyapunov technique [36], it can be proved that the gap between the average cost achieved by WACE and that by the optimal offline algorithm is bounded. The batch job queue is also upper bounded which translates into a finite queuing delay.

ALGORITHM 3 WACE

1: At the beginning of each time t, observe the DC state information $r(t)$, $\varepsilon_D(t)$, $\varepsilon_I(t)$, $\phi(t)$ and $p_B(t)$, for $t = 0,1,2,\ldots,T-1$

2: Choose $s(t)$ and $m(t)$ subject to (19), (20), (22) to minimize

$$V \cdot g(t) - q(t) \cdot b(t) \tag{24}$$

3: Update $q(t)$ according to (23)

4.4 Performance Evaluation of WACE

A trace simulation is performed to evaluate WACE. For details on simulation setups and data sets, please refer to Ref. [19]. Three benchmarks are provided for comparison with WAce:

- *SAVING* only considers electricity cost of the DC for optimization without any awareness on water and carbon.
- *CARBON* only optimizes the carbon emission of the DC without considering the impact of electricity and water.
- *ALWAYS* scheme has no optimization and it processes jobs as soon as possible.

In the first simulation, the cost–delay parameter V is fixed and comparison of the performance of WACE with three benchmark algorithms is depicted in Fig. 7. Figure 7A depicts that WACE achieves the lowest average total cost among all the algorithms. In Figure 7B showing average delay, WACE achieves a lower average total cost by opportunistically processing batch jobs

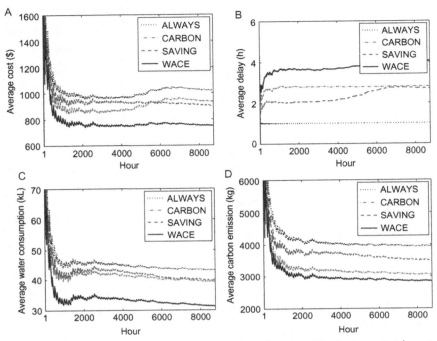

Figure 7 Comparison between WACE and benchmarks. (A) Average total cost, (B) average delay, (C) water consumption, and (D) carbon emission.

when the combined cost factor is relatively lower. The water consumption and carbon emission results in Fig. 7C and D display that compared to WACE, the benchmark algorithms consume more water and have higher carbon emission.

Figure 8A displays that the average electricity consumption remains almost same with varying water and carbon weights. The reason is that the actual energy consumption for processing a fixed amount of workloads remains relatively the same. In Fig. 8B, it can be seen that increase in either water or carbon weight increases the electricity cost. With increased water and/or carbon weight, WACE schedules batch jobs to find low water consumption and/or carbon emission due to sustainability considerations. Figure 8C and D depicts the decreasing trend of water consumption and carbon emission as the corresponding weighting factor is increased. The increased weighting factor means a higher priority in the optimization algorithm.

In Fig. 9A, the average total cost decreases for all algorithms with increased delay constraint. Particularly, WACE has the lowest average total

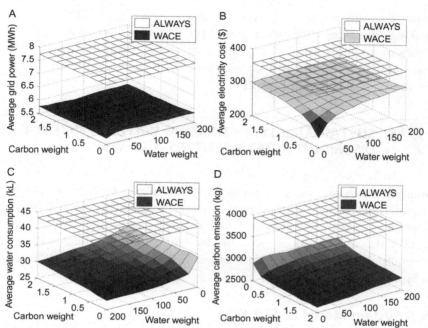

Figure 8 Impact of water and carbon weights. (A) Average electricity consumption, (B) electricity cost, (C) water consumption, and (D) carbon emission.

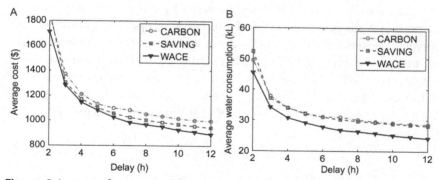

Figure 9 Impact of average delay constraint. (A) Average cost and (B) water consumption.

cost given any delay constraint. The performance gap between WACE and other two algorithms (i.e., SAVING and CARBON) increases with more relaxed delay constraint. Figure 9B displays a similar pattern of the change in water consumption with varying delay constraints. WACE has the lowest water consumption as it incorporates time-varying water efficiency when making scheduling decisions. Figure 9 presents an important guidance for choosing an appropriate set of water and carbon weights so that the DC can run in low cost and/or reduced water footprints without much impact on the average delay performance.

5. SUSTAINABILITY: OPTIMIZING WATER EFFICIENCY IN DISTRIBUTED DATA CENTERS (GLB-WS)

In Section 3, we discussed about the importance of water and sustainability and in Section 4, we presented a system management technique that exploits only the temporal diversity of water. In this section, we will present a system management technique which simultaneously considers both temporal and spatial diversities. We consider a number of distributed DCs and formulated the optimization problem as maximizing the water efficiency subject to budget constraints.

5.1 System Model of GLB-WS

We consider a discrete-time model with equal-length time slots indexed by $t = 1,2,\dots$ As in Refs. [4, 7], the focus is on hour-ahead prediction and hourly decisions. There are N geo-distributed DCs, indexed by $i = 1,2,\dots,N$. Each DC is partially powered by on-site renewable energy plants (e.g., solar panel and/or wind turbines) and contains M_i homogeneous servers. The service rate (defined as the *average* number of jobs processed in a unit time) of a server in DC i is μ_i. Let $m_i(t) \in [0, M_i]$ denote the number of active servers in DC i at time t, which can be relaxed into positive noninteger values as in Refs. [6, 7]. The focus is on server power consumption and the total server power consumption of DC i at time t can be expressed as

$$p_i\big(a_i(t), m_i(t)\big) = m_i(t) \cdot \left[e_{0,i} + e_{c,i}\frac{a_i(t)}{m_i(t)\mu_i}\right], \qquad (25)$$

where $a_i(t) = \sum_{j=1}^{J}\lambda_{i,j}(t)$ is the total amount of workloads dispatched to DC i, $_{0,i}$ is the static server power regardless of the workloads, and $e_{c,i}$ is the computing power incurred only when a server is processing workloads in DC i.

The electricity price in DC i at time t, denoted by $u_i(t)$, is known to the DC operator no later than the beginning of time t. $u_i(t)$ may change over time if the DCs participate in real-time electricity markets (e.g., hourly market [7]). Given the amount of available on-site renewable energy, $r_i(t) \in [0, r_{i, \max}]$, the incurred electricity cost of DC i can be expressed as

$$e_i(a_i(t), m_i(t)) = u_i(t)[\gamma_i(t) \cdot p_i(a_i(t), m_i(t)) - r_i(t)]^+, \qquad (26)$$

where $\gamma_i(t)$ is the PUE of DC i and $[\cdot]^+ = \max\{\cdot, 0\}$ indicates that no electricity will be drawn from the power grid if on-site renewable energy is already sufficient. As in Refs. [4, 7], Eq. (26) represents a linear electricity cost function; it can be extended to nonlinear convex functions (e.g., DCs are charged at a higher price if it consumes more power). The direct water consumption is obtained by multiplying the server power consumption with direct WUE, whereas the indirect water consumption depends on the electricity usage and the local EWIF. Thus, the water consumption of DC i at time t can be expressed as

$$w_i(t) = \epsilon_{i,d}(t) \cdot p_i(a_i(t), m_i(t)) + \epsilon_{i,id}(t) \cdot [\gamma_i(t) \cdot p_i(a_i(t), m_i(t)) - r_i(t)]^+, \qquad (27)$$

where $\epsilon_{i,d}(t)$ is the direct WUE at time t and $\epsilon_{i,id}(t)$ is the EWIF of the electricity powering DC i.

As in Refs. [4, 7], there are J gateways, each of which represents a geographically concentrated source of workloads (e.g., a state or province). The incoming workloads are forwarded to the N geo-distributed DCs. Let $\lambda_j(t) = [0, \lambda_{j, \max}]$ denote the workload arrival rate at the jth gateway. The workload is dispatched to DC i at a rate of $\lambda_{i,j}(t)$. As in Refs. [5, 7, 68], $\lambda_j(t)$ is assumed to be available (e.g., by using regression-based prediction) at the beginning of each time slot t. By incorporating the network transmission delay, the end-to-end average delay of workloads scheduled from gateway j to DC i is

$$d_{i,j}(a_i(t), m_i(t)) = \frac{1}{\mu_i - a_i(t)/m_i(t)} + l_{i,j}(t), \qquad (28)$$

where $a_i(t) = \sum_{j=1}^{J} \lambda_{i,j}(t)$ represents the total workloads processed in DC i, and $l_{i,j}(t)$ is average network delay approximated in proportion to the distance between DC i and the jth gateway, which can be well estimated by various approaches such as mapping and synthetic coordinate approaches. Intuitively, $d_{i,j}(a_i(t), m_i(t))$ is increasing in $a_i(t)$ and decreasing in $m_i(t)$ [4, 6, 7]. For example, the service process at each server can be modeled as an M/G/1/PS queue [7].

5.2 Problem Formulation of GLB-WS

The goal is to maximize the over water efficiency of distributed DCs given a budget constraint [20]. Based on the metric developed for measuring DC water efficiency [50, 57], the focus is on maximizing the overall (hourly) WUE of all the DCs, specified as follows

$$g(\lambda(t), \mathbf{m}(t)) = \frac{\sum_{i=1}^{N} w_i(t)}{\sum_{i=1}^{N} p_i(a_i(t), m_i(t))} \tag{29}$$

where $w_i(t)$ is the water usage and $p_i(a_i(t), m_i(t))$ is the server power consumption in DC i, given by (27) and (25), respectively.

The advantages for maximizing WUE are as follows: First, optimizing hourly WUE [50, 57] provides more timely information and may facilitate DC managers to improve their operations more promptly. Second, alternative objectives such as (weighted) sum WUE of individual DCs and total water consumption can also be optimized as variants of GLB-WS [20]. Third, one of the constraints is imposed on the total electricity cost (i.e., electricity budget), which implicitly bounds the maximum electricity consumption.

The optimization problem formulation is given as:

$$\mathbf{P6}: \quad \underset{\mathcal{A}(t)}{\text{maximize:}} \quad g(\lambda(t), \mathbf{m}(t)) \tag{30}$$

subject to:

$$\text{Maximum average delay:} \quad d_{i,j}(a_i, m_i) \leq D, \quad \forall i, j, t, \tag{31}$$

$$\text{Imposed capacity:} \quad m_i(t) \leq M_i, \quad \forall i, t, \tag{32}$$

$$\text{Workload dropping:} \quad \sum_{i=1}^{N} \lambda_{i,j}(t) = \lambda_j(t), \quad \forall j, t, \tag{33}$$

$$\text{Sever overloading:} \quad m_i(t)\mu_i > a_i(t) = \sum_{j=1}^{J} \lambda_{i,j}(t), \tag{34}$$

$$\text{Electricity budget:} \quad \sum_{i=1}^{N} e_i(a_i(t), m_i(t)) \leq B(t), \quad \forall i, t, \tag{35}$$

where $e_i(a_i, m_i)$ is the electricity cost of DC i and $B(t)$ is treated as exogenously given. $\mathcal{A}$ represents the server provisioning and load distribution decisions, i.e., $\mathbf{m}(t)$ and $\lambda(t)$, which need to be optimized. Solving the problem $\mathbf{P6}$ only requires the current electricity price, incoming workloads, available on-site renewable energies, and local WUEs, which are readily available in practice by leveraging hour-ahead prediction techniques [4, 6, 7].

5.3 Scheduling Algorithm (GLB-WS)

The problem **P6** is reformulated as linear-fractional programming [24]. The nonlinear delay constraint in (31) and the operator $[\cdot]^+ = \max\{\cdot, 0\}$ prohibit direct application of linear-fractional optimization. This difficulty is circumvented by rewriting (31) as $\sum_{j=1}^{J} \lambda_{i,j}(t) \leq (\mu_i - \frac{1}{D}) m_i(t)$, $\forall i, t$ and by introducing an auxiliary decision variable $z_i(t)$ indicating the amount of electricity usage of DC i such that $z_i(t) \geq p_i(a_i(t), m_i(t)) - r_i(t)$ and $z_i(t) \geq 0$. Thus, the linear-fractional programming problem can be expressed as:

$$\mathbf{P6'}: \quad \underset{\mathcal{A} \cup z}{\text{minimize:}} \quad \frac{\sum_{i=1}^{N} [\epsilon_{i,d} \cdot p_i(a_i, m_i) + \epsilon_{i,id} \cdot z]}{\sum_{i=1}^{N} p_i(a_i, m_i)} \tag{36}$$

$$\text{subject to :}$$
$$\text{constraints } (32) - (35) \tag{37}$$

$$\sum_{j=1}^{J} \lambda_{i,j} \leq \left(\mu_i - \frac{1}{D + l_{i,j}} \right) m_i, \quad \forall i, j, \tag{38}$$

$$z_i \geq 0 \text{ and } z_i \geq p_i(a_i, m_i) - r_i, \quad \forall i, \tag{39}$$

where, for brevity, the time index t is omitted without causing ambiguity.

From (25), it can be seen that server power $p_i(a_i(t), m_i(t))$ is affine in $\mathbf{m}(t)$ and $\lambda(t)$. Therefore, **P6'** belongs to linear-fractional programming (and also quasi-convex programming) [24]. An efficient iterative algorithm, called GLB-WS (Geographical Load Balancing for Water Sustainability), which is based on bisection method to solve **P6'**, was proposed. Another auxiliary variable $v \geq 0$ is introduced and the following inequality is defined

$$\sum_{i=1}^{N} [\epsilon_{i,d} \cdot p_i(a_i, m_i) + \epsilon_{i,id} \cdot z_i] \leq v \cdot \sum_{i=1}^{N} p_i(a_i, m_i). \tag{40}$$

Then, the bisection-based iterative method is described in Algorithm 4, where *MaxNum* is the maximum possible WUE and $\epsilon > 0$ is a small positive number governing the stopping criterion. At each iteration, a feasibility checking problem is solved, which is linear programming and hence easy to solve [24]. The final output of Algorithm 4 is a feasible decision satisfying (40) and constraints (37)–(39). The Algorithm 4 requires exactly $\lceil \log_2(\frac{ub - lb}{\epsilon}) \rceil$ iterations, and the final WUE will be within $\epsilon > 0$ of the optimum. Since the total complexity of solving Algorithm 4 is quite affordable for DCs (even for very small $\epsilon > 0$), it makes an appealing candidate for future geographic load balancing decisions.

ALGORITHM 4 GLB-WS

1: Input λ_j, r_i, $\epsilon_{i,d}$, $\epsilon_{i,id}$, and u_i, for $i = 1, 2, \ldots, N$
2: Initialize: $lb = 0$, $ub = MaxNum$, $v = \frac{lb+ub}{2}$
3: **while** $ub - lb > \epsilon$ **do**
4:    Check if there exist **m**, λ, and z that satisfy (40) and constraints (37)–(39);
5:    **if** "yes," **then** $ub = v$; **else** $lb = v$
6:    $v = lb + ub2$
7: **end while**

5.4 Performance Evaluation of GLB-WS

Trace-based simulations are performed to validate GLB-WS. For details on simulation setups and data sets, please refer to Ref. [20]. GLB for electricity Cost minimization [2, 4], called **GLB-Cost**, is chosen as a benchmark. As depicted in Fig. 10B, GLB-WS significantly improves the DC water efficiency compared to GLB-Cost. This is because GLB-WS can dynamically schedule workloads to water-efficient DCs, while GLB-Cost only considers cost effectiveness without taking water efficiency into account. A direct by-product of GLB-WS is the reduced water consumption. As shown in Fig. 10C, compared to GLB-Cost, GLB-WS reduces the water consumption substantially. The focus of GLB-WS is satisfying the given budget constraint. As can be seen from Fig. 10D, GLB-WS incurs a higher electricity bill compared to GLB-Cost. Note that the average electricity energy consumption by GLB-WS and GLB-Cost are almost the same. While electricity cost is certainly important for DCs, water sustainability is also critical and needs to be taken into consideration in the future design of DCs. As can be seen from Fig. 11B, it may not be possible to optimize water efficiency and cost effectiveness, simultaneously. An inherent trade-off exists between (water) sustainability and DC operational cost. GLB-WS is water-driven and can effectively schedule workloads to water-efficient DCs. Figure 11A depicts the average normalized WUE where DCs #3 and #4 are water efficient (but cost inefficient), while DCs #1 and #2 are on the opposite side. Thus, it can be observed from Fig. 11B that GLB-WS can schedule more workloads to DCs #3 and #4, whereas GLB-Cost favors DCs #1 and #2

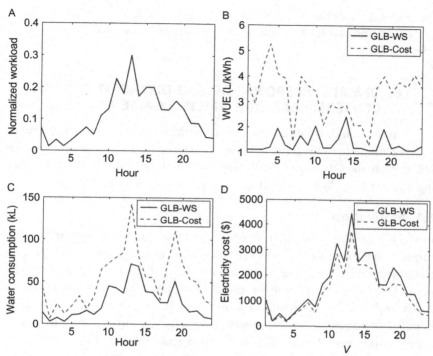

Figure 10 FIU workload trace and performance comparison between GLBWS and GLB-Cost. (A) FIU workload trace on May 1, 2012; (B) WUE comparison between GLB-WS and GLB-Cost; (C) water consumption comparison; and (D) electricity cost comparison.

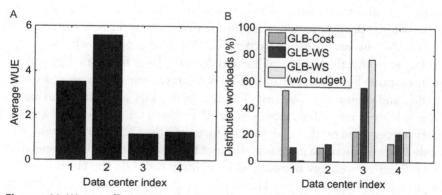

Figure 11 Water efficiency and distributed workloads in each data center. (A) Normalized WUE in each data center and (B) workload distribution under GLB-WS and GLB-Cost.

for processing workloads. As depicted in Fig. 11B, without considering budget constraint, GLB-WS will schedule almost all workloads to water-efficient DCs.

6. DEMAND RESPONSE OF GEO-DISTRIBUTED DATA CENTERS: REAL-TIME PRICING GAME APPROACH

In Section 2, we discussed cost minimization for DCs and in Sections 3–5, we studied sustainability in terms of water efficiency. In this section, we will explore the DR of geographically distributed DCs using a dynamic pricing scheme [22]. We will further study colocation DR in Section 7.

6.1 Motivation

First, DCs are large consumers of energy and their growing consumption has impacts on power prices. However, most of the existing research on DC cost minimization has made the obsolete assumption that *the electricity price applying on DC does not change with demand*. Second, DR is an important feature of smart grid [69]. By providing incentives, DR programs induce dynamic demand management of customers' electricity load in response to power supply conditions. Thus, DCs should participate in DR programs.

6.2 Challenge

The objective of DR programs is to reduce the peak-to-average load ratio. By using real-time pricing schemes, one DR program is persuading customers to shift their energy demand away from peak hours. The challenge is how to set the *right price* for the customer not only at the *right time* but also on the *right amount* of customers' demand. A real-time pricing scheme is regarded as an effective scheme if it can reduce the large fluctuation of energy consumption between peak and off-peak hours to increase power grid's reliability and robustness. The first challenge of this DR problem is the interaction between geo-distributed DCs and their local utilities. The utilities set their prices based on the total demand including the DCs' demand, which is only known when the price is available. On the other hand, DCs' operator will disperse its energy demand geographically based on the electric prices adjusted by the local utilities. The second challenge is the interaction among local utilities feeding power to the geo-distributed DCs. Especially, the local utilities set the electricity prices and the DCs' decisions are dependent on them. If any local utility changes its prices, other utilities are affected by it. In practice, the utilities are noncooperative, and thus, there exists a bottleneck on how to design a pricing mechanism that can enable an equilibrium price profile. To overcome the two above discussed challenges, the geo-distributed DR problem is transformed into a two-stage Stackelberg game. In Stage I, each utility will set a real-time price to maximize its own profit. In Stage II, given these prices, the DCs' operator will minimize its cost via workload shifting and dynamic server allocation.

6.3 Related Work for Demand Response

The electricity price is assumed to be fixed in many existing research on DCs' cost minimization which does not mirror any DR programs [4, 7, 70]. Based on the interactions between DCs and utilities, existing works considering DR of geo-distributed DCs can be divided into two categories: one-way interaction [71, 72] and two-way interaction [73, 74]. In Ref. [71], the authors studied coincident peak pricing (CPP), one of the most popular DR programs of DCs. Nonetheless, current DCs have no active response to the warning signals because of the uncertainty of these warnings [71], which inspire researchers to devise more effective DR approaches. In Ref. [72], a "prediction-based" method is used where the customers respond to the prices which are chosen based on a supply function. In Refs. [73, 74], dynamic pricing mechanisms are considered that make utilities and DCs coupled. Since there is no information exchange between utilities in reality, the system model of [74] that assumes all utilities cooperate to solve a social optimization problem is not relevant to current practice. On the other hand, Ref. [73] is based on a heuristic approach, and its pricing scheme cannot maximize the utilities' profit as well as minimize their cost.

6.4 DCs' Cost Minimization in Stage II

Let $t \in \mathcal{T} = \{1, ..., T\}$ denotes a discrete time slot of a billing cycle (e.g., typically a month). Let $\mathcal{I} = \{1, ..., I\}$ denotes the set of sites where DCs are located. Each DC i is assumed to be powered by a local utility company and has S_i homogeneous servers.

There are typically two types of workloads: delay-sensitive nonflexible interactive jobs and delay-tolerant flexible batch jobs. As in Ref. [70], each DC processes its batch jobs locally. The total arrival rate of interactive jobs to a DSP at time t is denoted by $\Lambda(t)$. The DCs' front-end server splits the total incoming workload $\Lambda(t)$ into separate workloads of geo-dispersed DCs, denoted by $\{\lambda_i(t)\}_{i \in \mathcal{I}}$.

The DSP minimizes its energy and migration cost, while it guarantees the QoS requirements of the interactive jobs. Let $e_i^b(t)$ denotes the amount of energy that the batch job processing consumes of each DC i in time slot t. Furthermore, the energy consumption of delay-sensitive jobs at DC i is given as [2]:

$$
\begin{aligned}
e_i^d(t) &= s_i(t)(P_{idle} + (P_{peak} - P_{idle})U_i(t) + (PUE(t) - 1)P_{peak}) \\
&= a_i\lambda_i(t) + b_i(t)s_i(t), \quad \forall i \in \mathcal{I}, t \in \mathcal{T},
\end{aligned}
\tag{41}
$$

where $s_i(t)$ is the server count, μ_i is the service rate of a server, P_{peak} and P_{idle} are the server's peak and idle power, respectively, $U_i(t) = \frac{\lambda_i(t)}{s_i(t)\mu_i}$ is the average server utilization, and $PUE(t)$ is the power usage effectiveness measuring the energy (e.g., cooling) efficiency of the DC. Furthermore, $a_i = (P_{peak} - P_{idle})/\mu_i$ and $b_i(t) = P_{idle} + (PUE(t) - 1)P_{peak}$. Therefore, the total energy consumption is denoted by

$$
e_i(t) = e_i^d(t) + e_i^b(t),
\tag{42}
$$

and given a price $p_i(t)$ at time t, the energy cost of DC i is $e_i(t)p_i(t)$.

Migrating the workload from front-end server to geo-distributed DCs can be very costly. Therefore, the migration cost to DC i can be modeled as [22]

$$
\omega d_i c_i(\lambda_i),
\tag{43}
$$

where d_i is the transmission delay from front-end server to DC i, ω is a weight factor, and $c_i(\lambda_i)$ is a function assumed to be strictly increasing and convex. d_i can be assumed to be a constant since it is proportional to the distance, and thus, migrating more requests from the front-end server to a farther DC is more costly. As it is widely used to penalize the action in control theory, a quadratic function, $c_i(\lambda_i(t)) = \lambda_i(t)^2$, is chosen for analysis tractability.

Each delay-sensitive request imposes a maximum delay D_i that the DSP has to guarantee when shifting this request to DC i [22]. Thus, the QoS constraint in terms of delay guarantee can be described as

$$
\frac{1}{s_i(t)\mu_i - \lambda_i(t)} + d_i \leq D_i, \quad \forall i,
\tag{44}
$$

where $1/(s_i(t)\mu_i - \lambda_i(t))$ is the average delay time of a request processed in DC i with arrival rate $\lambda_i(t)$ and service rate $s_i(t)\mu_i$ by queuing theory.

The model's emphasis is on two key controlling "knobs" of DCs' cost minimization [22]: the workload shifting to DC $\lambda_i(t)$ and the number of active servers provisioned $s_i(t)$ at site i, $\forall i$. Then, the Stage-II DC cost minimization is described as

$$\textbf{DC:} \quad \text{minimize:} \quad \sum_{t=1}^{T}\sum_{i=1}^{I} e_i(t)p_i(t) + \omega d_i\lambda_i(t)^2 \tag{45}$$

subject to: constraints (41), (42), (44)

$$\text{No workload dropping:} \quad \sum_{i=1}^{I}\lambda_i(t) = \Lambda(t), \quad \forall t, \tag{46}$$

$$\text{\# of active servers:} \quad 0 \leq s_i(t) \leq S_i, \quad \forall i, t, \tag{47}$$

$$\text{DC capacity:} \quad 0 \leq \lambda_i(t) \leq s_i(t)\mu_i, \quad \forall i, t, \tag{48}$$

$$\text{Variables:} \quad s_i(t), \lambda_i(t), \quad \forall i, t, \tag{49}$$

6.5 Noncooperative Pricing Game in Stage I

In this stage, the utility's revenue and cost models are described to form the individual objective of each utility's profit maximization. Afterward, the noncooperative pricing game between utilities is formulated.

6.5.1 Utility Revenue's Model

Solving **DC** can obtain the optimal energy consumption of DCs at time t, and it depends on prices $p_i(t)$, $\forall i$, of all utilities. Let $e_i(p(t))$ denotes the corresponding optimal power demand, where $p(t) := \{p_i(t)\}_{i\in\mathcal{I}}$. Let p_i^l and p_i^u, $\forall i, t$, denote the upper and lower bounds of the real-time prices that are imposed. Furthermore, each utility has its own background load (e.g., residential demand). Let $B_i(p_i(t))$ denotes the background load of utility i, which also responds to the price and can be modeled as [22]

$$B_i(p(t)) = \begin{cases} B_i^l, & p_i(t) \leq p_i^l; \\ \alpha - \beta p_i(t), & p_i^l \leq p_i(t) \leq p_i^u; \\ B_i^u, & p_i(t) \geq p_i^u, \end{cases} \tag{50}$$

where B_i^l and B_i^u are the minimum of maximum background demands of site i due to the physical constraints of consumers. Depending on the total power requested by DCs and background's demands, the revenue of utility i at time t is given by

$$rev_i(p(t)) = (e_i(p(t)) + B_i(p_i(t)))p_i(t). \tag{51}$$

6.5.2 Utility Cost's Model

When each utility serves the customers' load, it incurs a cost. The utility's cost also increases with increasing load because of blackouts due to overload. Thus, the utility's cost can be modeled based on a widely used electric load index (ELI) as

$$cost_i(p(t)) = \gamma ELI = \gamma \left(\frac{e_i(p(t)) + B_i(p_i(t))}{C_i(t)} \right)^2 C_i(t),$$

where $C_i(t)$ is utility i capacity at time t, and γ reflects the weight of the cost. A high value of ELI informs the utility to spend more for stability investment [74].

6.5.3 Stage-I Pricing Game Formulation

In practice, there is usually no communication exchange between the geo-distributed utilities to optimize the social performance. Instead, each utility i at time slot t maximizes its own profit, which is defined as the difference between revenue and cost and given by

$$u_i(p_i(t), p_{-i}(t)) = rev_i(p(t)) - cost_i(p(t)), \tag{52}$$

where $p_{-i}(t)$ denotes the price vector of other utilities except i. The profit of each utility not only depends on its energy price but also on the others'. Thus, the Stage-I utility profit maximization game is described as follows

- *Players*: the utilities in the set $\mathcal{I}$;
- *Strategy*: $p_i^l \leq p_i(t) \leq p_i^u$, $\forall i \in (I), t \in \mathcal{T}$;
- *Payoff function*: $\sum_{t=1}^{T} u_i(p_i(t), p_{-i}(t))$, $\forall i \in \mathcal{I}$.

6.6 Two-Stage Stackelberg Game: Equilibria and Algorithm

6.6.1 Backward Induction Method

The Stage-II DCs' cost minimization can be separated into independent problems at each time slot t. Time dependence notation is dropped for ease of presentation since only a specific time period is considered. In this stage, DCs cooperate with each other to minimize the total cost. This is achieved by determining the workload allocation λ_i and the number of active servers s_i

at each DC i. The DCs' cost minimization is a convex optimization problem.

First, constraint (44) is active because otherwise the DSP can decrease its energy cost by reducing $s_i(t)$. Hence, the equivalent of (44) is given by

$$s_i(\lambda_i) = \left[\frac{1}{\mu_i} \left(\lambda_i + \widetilde{D}_i^{-1} \right) \right]_o^{S_i}, \tag{53}$$

where $[\cdot]_x^y$ is the projection onto the interval $[x,y]$ and $\widetilde{D}_i := D_i - d_i$. In reality, most DCs can have enough number of servers to serve all requests at the same time due to the illusion of infinite capacity of DCs [7]. Thus, $s_i(\lambda_i) = \frac{1}{\mu_i} \left(\lambda_i + \widetilde{D}_i^{-1} \right)$ is adopted in the sequel. By substituting $s_i(\lambda_i)$ into the objective of **DC**, an equivalent problem **DC**$'$ can be formulated as follows [22]

$$\mathbf{DC}': \quad \underset{\lambda}{\text{minimize:}} \quad \sum_{i=1}^{I} f_i(\lambda_i) \tag{54}$$

$$\text{subject to:} \quad \sum_{i=1}^{I} \lambda_i = \Lambda, \tag{55}$$

$$\lambda_i \geq 0, \quad \forall i, \tag{56}$$

where $f_i(\lambda_i) := \omega d_i \lambda_i^2 + p_i \left(a_i + \frac{b_i}{\mu_i} \right) \lambda_i + p_i \left(e_b + \frac{b_i \widetilde{D}_i^{-1}}{\mu_i} \right)$. **DC**$'$ is a strictly convex problem that has a unique solution. DSP prefers to have $\lambda_i > 0$, $\forall i$ in order to utilize all DCs resources. The unique solution of **DC** and a necessary condition to achieve this solution with the optimal $\lambda_i^* > 0$, $\forall i$, can be characterized as follows:

Theorem 1. *Given a price vector* p, *the unique solutions of Stage-II* **DC** *problem are given as follows*

$$\lambda_i^* = \frac{\nu^* - p_i A_i}{2\omega d_i} > 0, \quad s_i^* = \frac{1}{\mu_i} \left(\lambda_i^* + \widetilde{D}_i^{-1} \right), \quad \forall i, \tag{57}$$

only if

$$\omega > \omega_{th}^l := \left(\hat{d} \max_i \{ p_i A_i \} - \sum_{i=1}^{I} p_i A_i / d_i \right) / 2\Lambda, \tag{58}$$

where $\hat{d} := \sum_{i=1}^{I} 1/d_i$, $A_i := a_i + \frac{b_i}{\mu_i}$ and $\nu^* = \frac{1}{\hat{d}}\left(2\omega\Lambda + \sum_{i=1}^{I} p_i A_i / d_i\right)$.

The condition (58) acts as a guideline for DSP in choosing an appropriate weight factor ω to ensure all DCs have positive requests.

Based on the Stage-II solutions, the Nash equilibrium of the Stage-I game can be characterized. In the noncooperative game, the focus is on whether a unique Nash equilibrium exists. If all other utilities' strategies p_{-i} are known, the best response strategy for utility i is given as

$$\mathbf{BR}_i(p_{-i}) = \arg\max_{p_i^l \leq p_i \leq p_i^u} u_i(p_i, p_{-i}), \quad \forall i, \tag{59}$$

where $u_i(p_i, p_{-i}) = \left(e_i^*(p) + B_i(p_i)\right)p_i - \gamma C_i\left(\dfrac{e_i^*(p) + B_i(p_i)}{C_i}\right)^2$ and $e_i^*(p) = (a_i\lambda_i^* + b_i s_i^*) + e_i^b$. With λ_i^* and s_i^* obtained from Theorem 1, $e_i^*(p)$ is equal to $\dfrac{A_i^2 p_i}{2\omega d_i}\left(\dfrac{1}{\hat{d} d_i} - 1\right) + \dfrac{A_i}{2\omega\hat{d} d_i}\sum_{j\neq i}\dfrac{A_j p_j}{d_j} + \dfrac{A_i\Lambda}{\hat{d} d_i} + \dfrac{b_i}{\mu_i\tilde{D}_i} + e_i^b$. p^e is a profile that satisfy $p_i^e = \mathbf{BR}_i(p_{-i}^e)$, $\forall i$. That is, at Nash equilibrium, every utility's strategy is its best response to others' strategies.

Theorem 2 (Existence and Uniqueness). *There exist a Nash equilibrium of the Stage-I game. Furthermore, if*

$$\omega \geq \omega_{th}^2 := \max_i \left[\frac{A_i\sum_{j\neq i}A_j/d_j - A_i^2\hat{d}(1 - 1/(d_i\hat{d}))}{2\beta\hat{d}d_i}\right], \tag{60}$$

then starting from any initial point, the best response strategies converge to a unique Nash equilibrium p^e of the Stage-I game.

6.6.2 Distributed Algorithm

A distributed algorithm [22], described in Algorithm 5, that can achieve the Nash equilibrium, was proposed. Algorithm 5 operates at the start of each pricing update period (i.e., 1 h). Until the algorithm converges to a price-setting equilibrium, it runs for many iterations. All the prices are collected from its local DCs by the front-end server which calculates the optimal energy consumption (line 4). Afterward, the front-end server will feedback these energy consumption data to its local DCs. The local DCs then forward their information to the local utility (line 5). Each utility finds an optimal price by solving its own profit maximization problem and then broadcasts this price to its local DCs and background customers (line 6).

The process is repeated until the prices converge to the Nash equilibrium (line 7). At this point, the final price is set and applied to the whole time slot t. According to Theorem 2, Algorithm 5 converges to a unique Nash equilibrium of Stage-I game.

ALGORITHM 5 Demand Response of Data Center with Real-Time Pricing

1: **initialize:** Set $k=0$, $p_i^{(0)}=p_i^u$, $\forall i$, and ω satisfies (60);
2: **repeat**
3:      Utility i broadcasts its $p_i^{(k)}$ to all customers, $\forall i$;
4:      The front-end server collects $p^{(k)}$ from all DCs, updates $e_i^*(p)^{(k)}$, and sends it back to DC i, $\forall i$;
5:      Each DC i reports its $e_i^*(p)^{(k)}$ to the local utility;
6:      Utility i receives the DRs from the local DC $e_i^*(pp^{(k)})$ and background users $B_i(p)^{(k)}$, and then solves $p_i^{(k+1)}=\mathbf{BR}_i(p_{-i}(k))$, $\forall i$;
7: **until** $|p^{(k+1)}-p^{(k)}| < \epsilon$.

First, it is assumed that the DSP deploys a front-end server for distribution of the incoming workload to DCs. In practice, a workload type can route only a subset of DCs due to the availability resource constraint of each DC. This issue can be addressed by incorporating more constraints into their model such as Ref. [14]. In practice, the workload types are classified at front-end server before routing. Second, it is assumed that DCs communicate with their front-end servers via one of the egress links of their Internet service provider (ISP). Specifically, the total time of one iteration includes the transmission time and the computational time. The transmission time from utilities to DCs (and vice versa) takes from 1 to 10 ms over a broadband speed of 100 Mbps, whereas it takes from 50 to 100 ms for a one-way communication between DCs and the front-end servers over a current ISP's path. The computational time is dependent on the processing power of the front-end server and smart meters in calculating (59), which is low-complexity problem and can be calculated in the timescale of microsecond [24].

6.7 Performance Evaluations

Trace-based simulations are conducted to verify the analysis and evaluate the performance of Algorithm 5. For details on simulation setups and data sets, please refer to Ref. [22]. Two baseline pricing schemes are considered for the simulation comparison as follows. The first baseline is based on the proposed dynamic pricing scheme of Ref. [73]. At each utility i, this pricing scheme is given as;

$$p_i(t+1) = \delta(PD_i(t) - PS_i(t)) + p_i(t), \tag{61}$$

where PD_i and PS_i are the power demand and supply of utility i at time t. For all simulation scenarios, $\delta = 0.5$. The second baseline is based on the Google's contract with their local utilities. According to Ref. [75], there are six Google's DCs powered by their local utilities at the following locations: The Dalles, OR; Council Bluffs, IA; Mayes County, OK; Lenoir, NC; Berkeley County, SC; and Douglas County, GA. In these locations, Google's DCs have long-term contracts with their local utilities with the following fixed rates [32.57, 42.73, 36.41, 40.68, 44.44, 39.97] \$/MWh, respectively.

Figure 12 depicts a sample-path optimal prices of three schemes at six locations corresponding to two workload traces. For Baseline 1 and Algorithm 5, the utilities' prices of these two schemes vary according to the workload pattern. The effect of migration cost to the optimal prices is also observed in Fig. 12. Since the nearest DCs to the front-end servers are sites 2 and 3, Fig. 12 shows that all dynamic pricing schemes set high prices at these sites compared with the other sites. It can also be seen that Baseline 1 always overprices Algorithm 5.

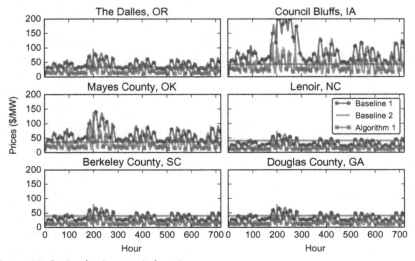

Figure 12 Optimal prices at six locations.

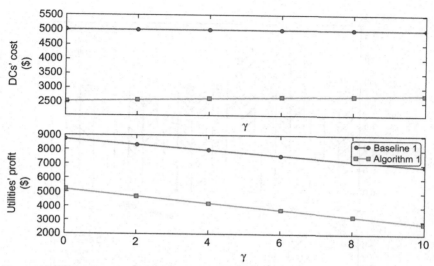

Figure 13 Effect of γ to average DCs' cost and utilities' profit.

In Fig. 13, the evaluation of the effect of parameter γ to average DCs' cost and utilities' profit is shown. First, Baseline 1 with higher prices has higher DCs' cost and utilities' profit than those of Algorithm 5. Thus, Algorithm 5 give more incentives to encourage the DCs to join the DR program. Second, when γ increases, the utilities' profit of both schemes decreases due to the cost in (52). In Algorithm 5, small γ is favorable because it can provide low DCs' cost and high utilities' profit.

PAR measures the effectiveness of designs for smart grid since the fluctuation of energy consumption between peak and off-peak hours indicates power grid's reliability and robustness. Reducing PAR is the ultimate goal of any DR program designs, and so is the goal of Algorithm 5. Figure 14 compares the PAR of three schemes with different γ. Figure 14 depicts that the PAR's performance of Algorithm 5 outperforms those of other schemes over time and space significantly.

7. DEMAND RESPONSE OF COLOCATION DATA CENTERS: A REVERSE AUCTION APPROACH

In Section 6, we studied DR of geographically distributed DCs. However, it does not fully cover DR for all DCs. In this section, we will investigate colocation DCs and focus on enabling DR for colocation DCs.

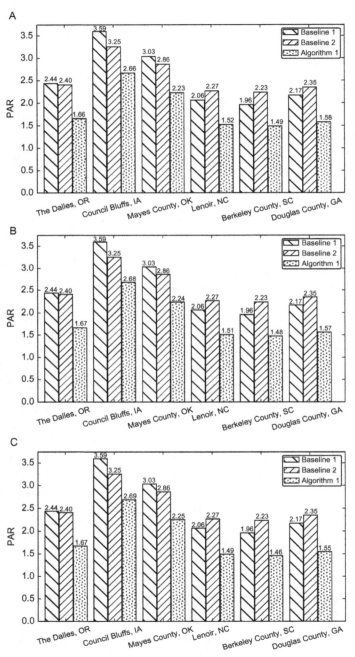

Figure 14 PAR at six locations. (A) $\gamma = 1$, (B) $\gamma = 4$, and (C) $\gamma = 8$.

7.1 Motivation and Challenge

DR program has been adopted as a national strategic plan for power grid innovation [69]. Reference [76] lists a comprehensive survey of various DR programs. By provisioning of economic incentives, DR can reshape customers' real-time electricity demand subject to time-varying supply availability [61, 77]. Due to their huge yet flexible energy demand, mega-scale DCs are ideal participants in DR programs [77–79]. Nevertheless, the existing efforts [77–82] have only focused on owner-operated DCs (e.g., Google and Amazon), without considering colocation DC. In owner-operated DC, the operator owns and has full control over the servers. On the other hand, colocation is a multitenant facility where multiple tenants put their own servers in one shared facility. The DC operator only provides reliable power supply, cooling, and network access.

A major challenge for colocation DR is **split incentive**. The colocation provider may desire DR for incentives from LSE. However, its tenants may not, as tenants are typically charged based on their subscribed peak power and there are no constraints on how much energy they consume or when they consume it [83].

In Ref. [21], "split incentive" hurdle for colocation DR is overcome by properly incentivizing tenants. Specifically, a novel market mechanism, based on reverse auction, which financially compensates tenants who are willing to reduce their energy consumption for DR, was proposed. The proposed mechanism is called iCODE (incentivizing COlocation tenants for DEmand response). It is fully voluntary and works in the following steps, as illustrated in Fig. 15. First, when utilities send DR signals/requests to colocation operator, they passed down to tenants who can submit bids. The bids include how much energy consumption they want to reduce and how much financial compensation they want to receive. Afterward, winning bids are selected by colocation operator so that the energy reduction can be maximized while the total payment to the tenants does not exceed that received from LSE. Finally, DR is executed according to the plan and payments are made to the tenants with the winning bids.

Two important reasons for colocation DCs are economic and efficiency. First, colocation is a necessary and important business model in DC industry. Some companies may not want to build their own DCs or completely outsource their IT requirements to public cloud providers (e.g., for economic or privacy concerns). Thus, colocation DC offers a "halfway" solution. Colocation tenants include not only small and medium businesses but also

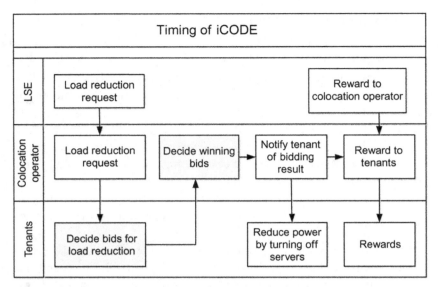

Figure 15 Timing of iCODE in colocation data center.

content distribution providers (e.g., Akamai) and many of the top-branded IT companies (e.g., Amazon and Microsoft) that wish to have global footprints for their last-mile service latency. Colocation DCs also host cloud computing services: e.g., medium-scale cloud providers, such as Salesforce and Box.com [84]. With the exponential growth of IT demands across all sectors, many colocation providers are also expanding their DC space [85]. In an analysis, colocation market is expected to grow at a compound annual growth rate of 11%, reaching US$ 43 billion by 2018 [86].

Second, colocations are more appropriate than owner-operated DCs for DR. Colocations demand a huge amount of power, and the peak power demand of colocations in New York region exceeds 400 MW (comparable to aggregate demand of Google's global DCs) [2, 87, 88]. An observation made from a Google study [89] is that "most large DCs are built to host servers from multiple companies" (i.e., colocations). Furthermore, many large colocations are often located in densely populated metropolitan areas (e.g., Los Angeles [88]) where DR is particularly desired for peak load shaving. On the other hand, most of the mega-scale owner-operated DCs (e.g., Google) are located in rural areas with very low population densities where the need for DR is less urgent.

Bringing unused servers off-line is one of the most extensively studied control knobs for energy saving [15, 89]. Switching servers between active and sleep/off modes can be easily automated [89]. Clearly, multiple tenants are required to cooperate via nontechnological mechanisms for DR. Market knobs, such as pricing and incentives, have been used to address various engineering issues [90, 91]. References [78, 79] show that owner-operated DCs are willing to shutdown some servers for DR incentives [78, 79].

Dynamically pricing energy usage for DR is a widely studied market mechanism [91]. However, in the context of colocation, it may not be plausible. First, directly "reselling" energy and adjusting energy price may be subject to strict government regulations [92]. Second, all tenants will face uncertain colocation costs, causing business reluctance and/or psychological concerns due to dynamic pricing [90, 93]. Finally, charging tenants to power utility's pricing is also not feasible, since tenants cannot plug their servers into utility's grid directly. Specifically, tenants need colocation operator's combined facility support (e.g., secured access, reliable power, cooling, network), not just facility space [94].

As illustrated in Fig. 15, a reverse auction-based incentive mechanism was proposed. In this context, "reverse" means that it is not the colocation operator who proactively offers rewards to tenants for energy reduction. Instead, tenants, at their own discretion, submit bidding information upon receiving a DR signal. iCODE [21] is "nonintrusive" and tenants are not enforced for DR or entitled any penalties if they do not participate in DR.

7.2 Related Work for Colocation Demand Response

We again emphasize that DCs are promising participants in DR programs. In-field test conducted shows that DCs can reduce energy consumption by 10–25% by participating in DR programs [77]. In Refs. [78, 80], authors studied resource management optimization for DR program and frequency regulation in power grid. Furthermore, Refs. [73, 74, 79] studied the interactions between the DCs and utilities, and pricing strategies by the utilities. However, these works only studied owner-operated DCs. In Ref. [81], authors address frequency regulation by controlling facility energy consumption via battery charging/discharging. However, this technique cannot scale with size because of limitations in battery size [95]. Thus, iCODE is the first study on aligning conflicting interests of colocation operator and tenants for colocation DR.

7.3 System Model

As in Refs. [77–79], time index is disregarded since the focus is on one-time DR, whose duration T is determined by LSE (e.g., 15 min to 1 h). Let N

denotes the number of tenants housing their servers in one colocation where tenant i owns M_i homogeneous servers. Tenants with heterogeneous servers can be easily viewed as multiple virtual tenants having homogeneous servers. For each server belonging to tenant i, let $p_{i,s}$ denotes a static/idle power, $p_{i,d}$ denotes dynamic power, and μ_i denotes service rate (measured in terms of the amount of workloads that may be processed in a unit time) [15]. During the DR period, the workload arrival rate, λ_i, can be predicted to a fairly reasonable accuracy using regression techniques [4, 15]. The baseline case is where no tenant participates in DR in which all servers are active and workloads are evenly distributed across servers for optimized performance. Thus, the average power consumption of tenant i's servers is

$p_i = M_i \cdot \left[p_{i,s} + p_{i,d} \cdot \dfrac{\lambda_i}{M_i \cdot \mu_i} \right] = M_i \cdot p_{i,s} + p_{i,d} \cdot \dfrac{\lambda_i}{\mu_i}$, where $\dfrac{\lambda_i}{M_i \cdot \mu_i}$ is the server

utilization. If tenant i decides to participate in DR by turning off $m_i \geq 0$ servers, then its average power will be $p_i' = (M_i - m_i) \cdot p_{i,s} + p_{i,d} \cdot \dfrac{\lambda_i}{mu_i}$. Hence, energy/load reduction by tenant i will be

$$\Delta e_i(m_i) = (p_i - p_i') \cdot T = m_i \cdot p_{i,s} \cdot T, \tag{62}$$

where $p_{i,s}$ is the static power and T is the DR duration.

There will be costs associated with turning off some servers, for example, switching cost and delay cost [15]. Bringing servers offline and then bringing them online back to normal operation incur switching/toggling costs, such as wear and tear [15]. Let α_i denote tenant i's switching cost for one server (quantified in monetary units), and therefore, the total switching cost for tenant i is $\alpha_i \cdot m_i$. The workload serving process at each server can be modeled as an M/M/1 queue. Thus, the average delay for tenant i's workload is

$1/\left(\mu_i - \dfrac{\lambda_i}{M_i - m_i} \right)$. The queuing model provides a reasonable approximation

for the actual service process [7, 13]. When the average delay exceeds a soft threshold $d_{i,th}$, the delay cost is incurred. The larger the soft delay threshold, the more delay-tolerant the tenant's workloads. The total delay cost can be

expressed as $d_i(m_i) = \lambda_i \cdot \left[\dfrac{1}{\mu_i - (\lambda_i/(M_i - m_i))} \right]^+$, where $[\cdot]^+ = \max\{0, \cdot\}$.

The colocation's non-IT energy reduction is captured by using the PUE factor γ, which typically ranges from 1.1 to 2.0 [89]. With a total IT energy reduction of $\sum_i \Delta e_i$ by tenants, the facility-level energy reduction will be $\gamma \cdot \sum_i \Delta e_i$. Let q denotes the price announced by LSE to procure a load reduction from customers. Then, the rewards provided by LSE to colocation operator are $q \cdot \gamma \cdot \sum_i \Delta e_i$.

7.4 Algorithm: iCODE

7.4.1 Deciding Tenants' Bids
For participation in DR, tenants need to be properly incentivized. Tenant i's requested payment for turning off m_i servers is given by

$$c_i(m_i) = w_i \cdot [\alpha_i \cdot m_i + \beta_i \cdot d_i(m_i)], \tag{63}$$

where $w_i \geq 1$ is referred to as *greediness* of tenant i, and $\beta_i \geq 0$ converts delay cost to monetary values [15]. Tenant i may submit multiple bids $(\Delta e_i, c_i)$, each corresponding to one value of $m_i \geq 0$. Furthermore, tenant i may only choose to turn off up to $-m_i$ servers so that the delay performance does not degrade below the agreed QoS. For convenience, denote the set of tenant i's bids as $\mathbf{b}_i \subseteq \mathbf{B}_i = \{(\Delta e_i, c_i) | (\Delta e_i(m_i), c_i(m_i)), m_i = 0, 1, ..., M_{i-1}\}$ such that $\mathbf{b}_i$ only contains valid bids.

7.4.2 Deciding Winning Bids
In Ref. [21], the objective is to maximize the total energy reduction subject to total compensation payment not exceeding that received from LSE. The problem of deciding winning bids (**DWB**) can be described as:

$$\mathbf{DWB}: \quad \underset{(\Delta e_i, c_i), \forall i \in \mathcal{I}}{\text{maximize:}} \quad \gamma \cdot \sum_{i \in \mathcal{I}} \Delta e_i \tag{64}$$

$$\text{subject to:} \quad \sum_{i \in \mathcal{I}} c_i \leq q \cdot \gamma \cdot \sum_{i \in \mathcal{I}} \Delta e_i, \tag{65}$$

$$(\Delta e_i, c_i) \in \mathbf{b}_i \cup \{(0,0)\}, \quad \forall i \in \mathcal{I}, \tag{66}$$

where $\mathcal{I}$ is the set of tenants who submit their bids to colocation operator, Eq. (64) specifies the objective of maximizing energy reduction, Eq. (65) indicates that the total compensation paid to tenants will not exceed the value received from the LSE, and Eq. (66) specifies that colocation operator can only choose "energy reduction, payment" pairs out of the bids submitted by tenants to honor their requests. $\{(0,0)\}$ is added in (66) to indicate that not all tenants' energy reduction requests will be accepted.

DWB is NP-hard but there exist various approximate solutions [96]. One approach is to solve **DWB** based on branch and bound technique that can yield a suboptimal solution with a reasonably low complexity [25]. Specifically, if e_i is allowed to take continuous values, then the requested payment in (63) is convex in e_i, and **DWB** becomes convex programming, for which there exists time-efficient methods [24]. The resulting energy reduction is an upper bound on the optimal value of **DWB**. On the other hand, if a greedy-based approach is chosen (e.g., select the bids in ascending order of $\Delta e_i/c_i$), then a lower bound on the optimal value of **DWB** will be obtained. For sufficiently close upper and lower bounds, the greedy solution can be chosen, due to the closeness of energy reduction to the upper bound (and hence optimum too). Otherwise, **DWB** can be recursively solved by fixing some bids to be selected and solving a smaller-scale subproblem. Lower/upper bounds are computed via greedy/relaxation approach to solve the subproblem. If the bounds remain still far apart, the subproblem is further decomposed into an even smaller-scale subproblem. This process is repeated until the gap between the two bounds are sufficiently small or the maximum iteration number is reached. Since colocation operator receives DR signal from LSE well beforehand and there is no need to solve **DWB** in real time, the computational complexity of solving **DWB** is not a major bottleneck [78, 79].

7.5 Performance Analysis

To demonstrate its effectiveness, the scenario in which no tenants participate in DR, called NDR (nondemand response), is chosen as the benchmark. Please refer to Ref. [21] for details on simulation settings. The comparison between iCODE and NDR is given in Fig. 16. First, Fig. 16A shows the energy reduction by iCODE compared to NDR. It can be seen that more than 4 MWh energy reduction per hour can be achieved, which is a fairly significant energy reduction. Next, Fig. 16B describes the hourly energy consumptions by iCODE and NDR, indicating that more than 50% energy can be slashed in some hours due to the low server utilization in colocation. Figure 16C displays the monetary incentive received by different tenants. There is some "residual" incentive paid to colocation operator by LSE, because sometimes tenants do not seek as high incentives as LSE provides. Figure 16D depicts the barely noticeable performance degradation experienced by tenants compared to their soft delay thresholds. There is an up to around 1.5 ms increase in average delay beyond the threshold for delay-sensitive tenants #1 and 0.5 ms for tenant #2. Tenant #3 has 30-s delay, which is acceptable for delay-tolerant workloads.

    To counter unexpected possible traffic spikes, tenants can deliberately overestimate the workload arrival rate by a certain overprediction factor

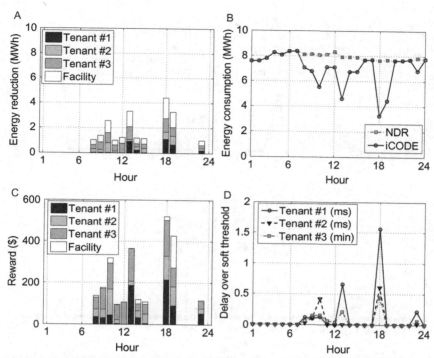

Figure 16 Comparison between iCODE and NDR. (A) Energy reduction by iCODE. (B) Energy consumption by iCODE and NDR. (C) Incentives received. No incentives are provided in NDR. (D) Average delay exceeding the soft threshold in iCODE.

$\phi \geq 0$: the higher ϕ, the more overpredicts. Intuitively, when tenants are more conservative and tend more to overpredict workloads, fewer number of servers will be turned off. However, Fig. 17A and B shows that even when tenants overestimate the workloads by 30%, the energy reduction for DR is not significantly compromised. The authors choose 30% because the workload prediction error is typically within 30% [63].

Tenants may desire more than their true costs for turning off servers. Thus, the greediness factor w_i for tenants is increased. Equivalently, this captures the scenarios where tenants are less willing to participate in DR unless they receive sufficiently large incentives. Figure 17C and D displays that as tenants are becoming more greedy, the performance becomes better and the energy reduction decreases. As seen by comparing Figs. 16D and 17D, asking for higher payments than actual costs may not be of tenants' interests, because doing so will reduce tenants' financial rewards yet without improving their delay performances.

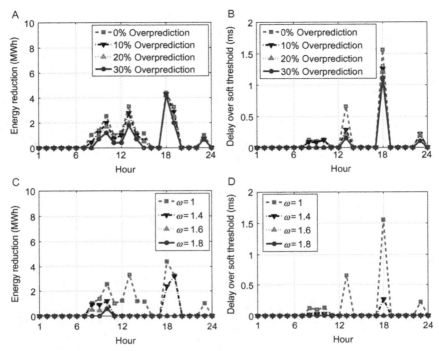

Figure 17 Impact of workload overprediction (A and B) and greediness (C and D). (A) Workload overprediction impact on energy reduction, (B) tenant #1's delay over soft threshold, (C) greediness impact on energy reduction, and (D) tenant #1's delay over soft threshold.

8. CONCLUSION AND OPEN CHALLENGES

We have presented some novel system management issues in DCs with three different perspectives: cost minimization, water sustainability, and DR. In Section 2, we discussed about cost minimization and presented novel management techniques. In Section 2.2, power management is performed for a fixed budget in a single DC. We employed Lyapunov optimization technique [36] to develop an online algorithm eBud to cap the long-term energy constraint. In Section 2.3, we considered thermal-aware scheduling for geographically distributed DCs. We also employ Lyapunov optimization technique [36] and develop an online algorithm called GreFar. In Sections 3–5, we investigated the water sustainability and proposed two algorithms to minimize water usage and maximize WUE. In Section 4, we consider a single DC and temporal diversity of water efficiency whereas in Section 5, we consider geo-distributed DCs with both temporal and spatial diversity. For the former, we employed Lyapunov optimization technique [36] and for the latter, we use convex optimization techniques [24]. In Sections 6 and 7, we explored DR program. In Section 6, DR program for smart gird is studied and the problem is formulated into a two-stage Stackelberg game. This formulation captures the different objectives of two players, the DCs and the utilities. The proposed algorithm leads to a unique Stackelberg equilibrium where a stable price is reached. In Section 7, we investigated colocation DR where the DC operators and tenants have split incentives. We employed reverse auction to properly incentivize the tenants to participate in the DR program.

DC research is a vibrant research area and there remains many unexplored topics. In the future, due to ubiquitous computing, there will be many interesting interdisciplinary research topics. Some open challenges for DC research can be classified into three different areas:

- *Cost*: These include the traditional optimization methods for cost minimization and efficiency maximization. Future research direction should be for distributed DCs instead of individual DCs. This area will include and benefit not only DSPs but also ISPs.

- *Sustainability*: Installation of green energy (i.e., solar) has been increasing in recent years. However, most (renewable) energy generated (e.g., solar) is time dependent. Geographical load balancing with respect to green energy generated is an interesting research topic (e.g., DC load following the sun across geo-distributed DCs). We must not forget water and carbon sustainability and take into consideration.

- *Demand response*: This is an interdisciplinary topic between smart grid and DC, and the focus is on the strategic interactions between producer and consumer. The end goal is the power proportionality and the environment is dynamic. The utilities and DCs can jointly optimize their performance by cooperation. The cooperation design (e.g., auction, bargaining) is an open research area.

ACKNOWLEDGMENTS

This work was supported in part by the U.S. NSF under grants CNS-1423137 and CNS-1551661.

REFERENCES

[1] J.G. Koomey, Growth in data center electricity use 2005 to 2010, 2011. http://www.analyticspress.com/datacenters.html.

[2] A. Qureshi, R. Weber, H. Balakrishnan, J. Guttag, B. Maggs, Cutting the electric bill for Internet-scale systems, ACM SIGCOMM Comput. Commun. Rev. 39 (4) (2009) 123, ISSN 01464833, http://dx.doi.org/10.1145/1594977.1592584.

[3] California ISO, http://www.caiso.com/Pages/default.aspx.

[4] L. Rao, X. Liu, L. Xie, W. Liu, Minimizing electricity cost: optimization of distributed Internet data centers in a multi-electricity-market environment, in: IEEE INFOCOM, ISBN: 978-1-4244-5836-3, ISSN 0743-166X 2010, pp. 1–9, http://dx.doi.org/10.1109/INFCOM.2010.5461933.

[5] K. Le, R. Bianchini, T.D. Nguyen, O. Bilgir, M. Martonosi, Capping the brown energy consumption of Internet services at low cost, in: IEEE International Conference on Green Computing, ISBN: 978-1-4244-7612-1, 2010, pp. 3–14, http://dx.doi.org/10.1109/GREENCOMP.2010.5598305.

[6] M. Lin, Z. Liu, A. Wierman, L.L.H. Andrew, Online algorithms for geographical load balancing, in: IEEE International Green Computing Conference, ISBN: 978-1-4673-2154-9, 2012, pp. 1–10, http://dx.doi.org/10.1109/IGCC.2012.6322266.

[7] Z. Liu, M. Lin, A. Wierman, S.H. Low, L.L.H. Andrew, Greening geographical load balancing, in: ACM SIGMETRICS, ISBN: 9781450308144, 2011, p. 233, New York. http://dx.doi.org/10.1145/1993744.1993767.

[8] Google Inc., Google's Green PPAs: what, how, and why, Tech. Rep., http://static.googleusercontent.com/external_content/untrusted_dlcp/cfz.cc/en/us/green/pdfs/renewable-energy.pdf.

[9] Microsoft Corporation, Becoming carbon neutral: how Microsoft is striving to become leaner, greener, and more accountable, 2012.

[10] Y. Zhang, Y. Wang, X. Wang, Electricity bill capping for cloud-scale data centers that impact the power markets, in: 41st IEEE International Conference on Parallel Processing, ISBN: 978-1-4673-2508-0, ISSN 0190-3918, 2012, pp. 440–449, http://dx.doi.org/10.1109/ICPP.2012.23.

[11] N. Buchbinder, N. Jain, I. Menache, Online job-migration for reducing the electricity bill in the cloud, in: ACM NETWORKING, ISBN: 978-3-642-20756-3, 2011, pp. 172–185.

[12] B. Guenter, N. Jain, C. Williams, Managing cost, performance, and reliability tradeoffs for energy-aware server provisioning, in: IEEE INFOCOM, ISBN: 978-1-4244-9919-9, ISSN 0743-166X, 2011, pp. 1332–1340, http://dx.doi.org/10.1109/INFCOM.2011.5934917.

[13] A. Gandhi, M. Harchol-Balter, R. Das, C. Lefurgy, Optimal power allocation in server farms, ACM SIGMETRICS Perform. Eval. Rev. 37 (1) (2009) 157–168, ISSN 0163-5999, http://dx.doi.org/10.1145/2492101.1555368.

[14] P.X. Gao, A.R. Curtis, B. Wong, S. Keshav, It's not easy being green, in: ACM SIGCOMM, ISBN: 9781450314190, 2012, p. 211, New York. http://dx.doi.org/10.1145/2342356.2342398.

[15] M. Lin, A. Wierman, L.L.H. Andrew, E. Thereska, Dynamic right-sizing for power-proportional data centers, in: IEEE INFOCOM, 2011, pp. 1098–1106, http://dx.doi.org/10.1109/INFCOM.2011.5934885.

[16] L. Li, C.-J.M. Liang, J. Liu, S. Nath, A. Terzis, C. Faloutsos, ThermoCast: a cyber-physical forecasting model for datacenters, in: 17th ACM International Conference on Knowledge Discovery and Data Mining, ISBN: 978-1-4503-0813-7 2011, pp. 1370–1378, New York. http://dx.doi.org/10.1145/2020408.2020611.

[17] L. Parolini, N. Tolia, B. Sinopoli, B.H. Krogh, A cyber-physical systems approach to energy management in data centers, in: 1st ACM/IEEE International Conference on

Cyber-Physical Systems, ISBN: 978-1-4503-0066-7, 2010, pp. 168–177, New York. http://dx.doi.org/10.1145/1795194.1795218.

[18] M. Polverini, A. Cianfrani, S. Ren, A.V. Vasilakos, Thermal-aware scheduling of batch jobs in geographically distributed data centers, IEEE Trans. Cloud Comput. 2 (1) (2014) 71–84.

[19] M.A. Islam, K. Ahmed, S. Ren, G. Quan, Exploiting temporal diversity of water efficiency to make data center less "thirsty, in: 11th USENIX International Conference on Autonomic Computing, ISBN: 9781931971119, 2014.

[20] S. Ren, Optimizing water efficiency in distributed data centers, in: IEEE International Conference on Cloud and Green Computing, ISBN: 978-0-7695-5114-2, 2013, pp. 68–75, http://dx.doi.org/10.1109/CGC.2013.19.

[21] S. Ren, M.A. Islam, Colocation demand response: why do I turn off my servers? in: 11th IEEE International Conference on Autonomic Computing, ISBN: 9781931971119.

[22] N. Tran, S. Ren, Z. Han, S.M. Jang, Demand response of data centers: a real-time pricing game between utilities in smart grid, in: 9th USENIX International Workshop on Feedback Computing, 2014.

[23] M.A. Islam, S. Ren, G. Quan, Online energy budgeting for virtualized data centers, in: 21st IEEE International Symposium on Modelling, Analysis and Simulation of Computer and Telecommunication Systems, ISBN: 978-0-7695-5102-9, ISSN 1526-7539, 2013, pp. 424–433, http://dx.doi.org/10.1109/MASCOTS.2013.64.

[24] S. Boyd, L. Vandenberghe, Convex Optimization, Cambridge University Press, New York, NY, 2004, ISBN: 978-0-521-83378-3. http://dx.doi.org/10.1080/10556781003625177.

[25] S. Boyd, A. Ghosh, A. Magnani, Branch and bound methods, Stanford University, 2003.

[26] Q. Zhu, G. Agrawal, Resource provisioning with budget constraints for adaptive applications in cloud environments, IEEE Trans. Serv. Comput. 5 (4) (2012) 497–511, ISSN 1939-1374, http://dx.doi.org/10.1109/TSC.2011.61.

[27] S. Chaisiri, B.-S. Lee, D. Niyato, Optimization of resource provisioning cost in cloud computing, IEEE Trans. Serv. Comput. 5 (2) (2012) 164–177, ISSN 1939-1374, http://dx.doi.org/10.1109/TSC.2011.7.

[28] R. Urgaonkar, U.C. Kozat, K. Igarashi, M.J. Neely, Dynamic resource allocation and power management in virtualized data centers, in: IEEE Network Operations and Management Symposium, ISBN: 978-1-4244-5366-5, ISSN 1542-1201, 2010, pp. 479–486, http://dx.doi.org/10.1109/NOMS.2010.5488484.

[29] D. Ardagna, B. Panicucci, M. Trubian, L. Zhang, Energy-aware autonomic resource allocation in multitier virtualized environments, IEEE Trans. Serv. Comput. 5 (1) (2012) 2–19, ISSN 1939-1374, http://dx.doi.org/10.1109/TSC.2010.42.

[30] V. Shrivastava, P. Zerfos, K.-W. Lee, H. Jamjoom, Y.-H. Liu, S. Banerjee, Application-aware virtual machine migration in data centers, in: IEEE INFOCOM, ISBN: 978-1-4244-9919-9, ISSN 0743-166X, 2011, pp. 66–70, http://dx.doi.org/10.1109/INFCOM.2011.5935247.

[31] J. Xu, J.A.B. Fortes, Multi-objective virtual machine placement in virtualized data center environments, in: IEEE/ACM International Conference on Green Computing and Communications & International Conference on Cyber, Physical and Social Computing, ISBN: 978-1-4244-9779-9, 2010, pp. 179–188, http://dx.doi.org/10.1109/GreenCom-CPSCom.2010.137.

[32] H. Liu, C.-Z. Xu, H. Jin, J. Gong, X. Liao, Performance and energy modeling for live migration of virtual machines, in: 20th ACM International Symposium on High Performance Distributed Computing, ISBN: 9781450305525, 2011, p. 171, New York. http://dx.doi.org/10.1145/1996130.1996154.

[33] A.S.M.H. Mahmud, S. Ren, Online resource management for data center with energy capping, in: USENIX 8th International Workshop on Feedback Computing, 2013, San Jose, CA.

[34] H. Lim, A. Kansal, J. Liu, Power budgeting for virtualized data centers, in: USENIX Annual Technical Conference, 2011, p. 5.

[35] C. Ren, D. Wang, B. Urgaonkar, A. Sivasubramaniam, Carbon-aware energy capacity planning for datacenters, in: IEEE 20th International Symposium on Modeling, Analysis and Simulation of Computer and Telecommunication Systems, ISBN: 978-1-4673-2453-3, ISSN 1526-7539, 2012, pp. 391–400, http://dx.doi.org/10.1109/MASCOTS.2012.51.

[36] M.J. Neely, Stochastic network optimization with application to communication and queueing systems, Synth. Lect. Commun. Netw. 3 (1) (2010) 1–211, ISSN 1935-4185, http://dx.doi.org/10.2200/S00271ED1V01Y201006CNT007.

[37] M. A. Islam, S. Ren, H. Mahmud, G. Quan, Online energy budgeting for cost minimization in virtualized data center, supplementary materials, http://users.cis.fiu.edu/~sren/doc/tech/mascots_2013_full.pdf.

[38] X. Chen, X. Liu, S. Wang, X.-W. Chang, TailCon: power-minimizing tail percentile control of response time in server clusters, in: 31st IEEE Symposium on Reliable Distributed Systems, ISBN: 978-1-4673-2397-0, ISSN 1060-9857, 2012, pp. 61–70, http://dx.doi.org/10.1109/SRDS.2012.72.

[39] J. Moore, J. Chase, P. Ranganathan, R. Sharma, Making scheduling "cool": temperature-aware workload placement in data centers, in: USENIX Annual Technical Conference, 2005, p. 5. Berkeley, CA.

[40] S. Li, T. Abdelzaher, M. Yuan, TAPA: temperature aware power allocation in data center with Map-Reduce, in: IEEE International Green Computing Conference and Workshops, 2011, pp. 1–8, http://dx.doi.org/10.1109/IGCC.2011.6008602.

[41] J. Chen, R. Tan, Y. Wang, G. Xing, X. Wang, X. Wang, B. Punch, D. Colbry, A high-fidelity temperature distribution forecasting system for data centers, in: 33rd IEEE Real-Time Systems Symposium, 2012, pp. 215–224, http://dx.doi.org/10.1109/RTSS.2012.73.

[42] A. Beitelmal, C. Patel, Thermo-fluids provisioning of a high performance high density data center, Distrib. Parallel Databases 21 (2–3) (2007) 227–238, ISSN 0926-8782, http://dx.doi.org/10.1007/s10619-005-0413-0.

[43] B. Hayes, Cloud computing, ACM Commun. 51 (7) (2008) 9–11, ISSN 0001-0782, http://dx.doi.org/10.1145/1364782.1364786.

[44] T. Mukherjee, A. Banerjee, G. Varsamopoulos, S.K.S. Gupta, S. Rungta, Spatio-temporal thermal-aware job scheduling to minimize energy consumption in virtualized heterogeneous data centers, Comput. Netw. 53 (17) (2009) 2888–2904, ISSN 1389-1286, http://dx.doi.org/10.1016/j.comnet.2009.06.008.

[45] T. Mukherjee, Q. Tang, C. Ziesman, S.K.S. Gupta, P. Cayton, Software architecture for dynamic thermal management in datacenters, in: 2nd International Conference on Communication Systems Software and Middleware, 2007, pp. 1–11, http://dx.doi.org/10.1109/COMSWA.2007.382430.

[46] K. Mukherjee, S. Khuller, A. Deshpande, Saving on cooling: the thermal scheduling problem, in: 12th ACM SIGMETRICS/PERFORMANCE, ISBN: 978-1-4503-1097-0, 2012, pp. 397–398, New York. http://dx.doi.org/10.1145/2254756.2254811.

[47] S. Ren, Y. He, F. Xu, Provably-efficient job scheduling for energy and fairness in geographically distributed data centers, in: IEEE 32nd International Conference on Distributed Computing Systems, 2012, pp. 22–31, http://dx.doi.org/10.1109/ICDCS.2012.77.

[48] AT&T, Water—AT&T 2012 sustainability report, http://www.att.com/gen/landing-pages?pid=24188.

[49] U.S Department of Energy, Energy demands on water resources, 2006. http://www.sandia.gov/energy-water/docs/121-RptToCongress-EWwEIAcomments-FINAL.pdf.

[50] The Green Grid, Water usage effectiveness (WUE): a Green Grid data center sustainability metric., 2011.

[51] M. J. Rutberg, Modeling water use at thermoelectric power plants., 2012.

[52] Uptime Institute, 2014 Data center industry survey results, Uptime Institute, 2014. http://symposium.uptimeinstitute.com/images/stories/symposium2014/presentations/mattstansberry-surveyresults2014.pdf.

[53] U.S. Green Building Council, Leadership in energy & environmental design, http://www.usgbc.org/leed.

[54] E. Frachtenberg, Holistic datacenter design in the open compute project, Computer 45 (7) (2012) 83–85, ISSN 0018-9162, http://dx.doi.org/10.1109/MC.2012.235.

[55] C. Bash, T. Cader, Y. Chen, D. Gmach, R. Kaufman, D. Milojicic, A. Shah, P. Sharma, Cloud sustainability dashboard dynamically assessing sustainability of data centers and clouds, HP Labs, 2011.

[56] R. Sharma, A. Shah, C. Bash, T. Christian, C. Patel, Water efficiency management in datacenters: metrics and methodology, in: IEEE International Symposium on Sustainable Systems and Technology, 2009, pp. 1–6, http://dx.doi.org/10.1109/ISSST.2009.5156773.

[57] Facebook, Prineville—Facebook Power, https://www.fbpuewue.com/prineville.

[58] Google Inc., Data center efficiency—data centers—Google, http://www.google.com/about/datacenters/efficiency/.

[59] S. Ren, Batch job scheduling for reducing water footprints in data center, in: 51st Annual Allerton Conference on Communication, Control, and Computing, 2013, pp. 747–754, http://dx.doi.org/10.1109/Allerton.2013.6736599.

[60] eBay, Digital service efficiency, http://tech.ebay.com/dashboard.

[61] U.S. Department of Energy, Open Data Catalogue. http://energy.gov/data/downloads/open-data-catalogue.

[62] J. Macknick, R. Newmark, G. Heath, K.C. Hallett, Operational water consumption and withdrawal factors for electricity generating technologies: a review of existing literature, Environ. Res. Lett. 7 (4) (2012). http://dx.doi.org/10.1088/1748-9326/7/4/045802.

[63] Z. Liu, Y. Chen, C. Bash, A. Wierman, D. Gmach, Z. Wang, M. Marwah, C. Hyser, Renewable and cooling aware workload management for sustainable data centers, in: 12th ACM SIGMETRICS/PERFORMANCE, ISBN: 978-1-4503-1097-0, 2012, pp. 175–186, New York. http://dx.doi.org/10.1145/2254756.2254779.

[64] R. Urgaonkar, B. Urgaonkar, M.J. Neely, A. Sivasubramaniam, Optimal power cost management using stored energy in data centers, in: ACM SIGMETRICS, ISBN: 978-1-4503-0814-4, 2011, pp. 221–232, New York. http://dx.doi.org/10.1145/1993744.1993766.

[65] Apple Inc., Apple—environmental responsibility, http://www.apple.com/environment/.

[66] J.R. Lorch, A.J. Smith, Improving dynamic voltage scaling algorithms with PACE, in: ACM SIGMETRICS, ISBN: 1-58113-334-0, 2001, pp. 50–61, New York. http://dx.doi.org/10.1145/378420.378429.

[67] X. Fan, W.-D. Weber, L.A. Barroso, Power provisioning for a warehouse-sized computer, ACM SIGARCH Comput. Archit. News 35 (2) (2007) 13–23, ISSN 0163-5964, http://dx.doi.org/10.1145/1273440.1250665.

[68] N. Deng, C. Stewart, D. Gmach, M. Arlitt, J. Kelley, Adaptive green hosting, in: 9th IEEE International Conference on Autonomic Computing, ISBN: 978-1-4503-1520-3, 2012, pp. 135–144, New York. http://dx.doi.org/10.1145/2371536.2371561.

[69] U.S. Federal Energy Regulatory Commission, Assessment of demand response and advanced metering, 2012.

[70] S. Ren, Y. He, COCA: online distributed resource management for cost minimization and carbon neutrality in data centers, in: IEEE/ACM International Conference for High Performance Computing, Networking, Storage and Analysis, ISBN: 9781450323789, 2013, pp. 1–12, New York. http://dx.doi.org/10.1145/2503210. 2503248.

[71] Z. Liu, A. Wierman, Y. Chen, B. Razon, N. Chen, Data center demand response: avoiding the coincident peak via workload shifting and local generation, ACM SIG-METRICS Perform. Eval. Rev. 41 (1) (2013) 341, ISSN 01635999, http://dx.doi.org/10.1145/2494232.2465740.

[72] P. Wang, L. Rao, X. Liu, Y. Qi, D-Pro: dynamic data center operations with demand-responsive electricity prices in smart grid, IEEE Trans. Smart Grid 3 (4) (2012) 1743–1754, ISSN 1949-3053, http://dx.doi.org/10.1109/TSG.2012.2211386.

[73] Y. Li, D. Chiu, C. Liu, L.T.X. Phan, T. Gill, S. Aggarwal, Z. Zhang, B.T. Loo, D. Maier, B. McManus, Towards dynamic pricing-based collaborative optimizations for green data centers, in: 29th IEEE International Conference on Data Engineering Workshops, 2013, pp. 272–278, http://dx.doi.org/10.1109/ICDEW.2013.6547462.

[74] H. Wang, J. Huang, X. Lin, H. Mohsenian-Rad, Exploring smart grid and data center interactions for electric power load balancing, ACM SIGMETRICS Perform. Eval. Rev. 41 (3) (2014) 89–94, ISSN 01635999, http://dx.doi.org/10.1145/2567529. 2567556.

[75] H. Xu, B. Li, Reducing electricity demand charge for data centers with partial execution. in: 5th ACM International Conference on Future Energy Systems, ISBN: 9781450328197, 2014, pp. 51–61, New York. http://dx.doi.org/10.1145/2602044. 2602048.

[76] The White House, U.S. Federal Leadership in Environmental, Energy and Economic Performance—executive order 13514, The White House, 2009. http://www. whitehouse.gov/assets/documents/2009fedleader_eo_rel.pdf.

[77] G. Ghatikar, V. Ganti, N. Matson, M.A. Piette, Demand response opportunities and enabling technologies for data centers: findings from field studies, 2012.

[78] M. Ghamkhari, H. Mohsenian-Rad, Energy and performance management of green data centers: a profit maximization approach. IEEE Trans. Smart Grid 4 (2) (2013) 1017–1025, ISSN 1949-3053, http://dx.doi.org/10.1109/TSG.2013.2237929.

[79] Z. Liu, I. Liu, S. Low, A. Wierman, Pricing data center demand response, in: ACM SIGMETRICS, ISBN: 978-1-4503-2789-3, 2014, pp. 111–123, New York. http://dx.doi.org/10.1145/2591971.2592004.

[80] D. Aikema, R. Simmonds, H. Zareipour, Data centres in the ancillary services market, in: IEEE International Green Computing Conference, 2012, pp. 1–10, http://dx.doi.org/10.1109/IGCC.2012.6322252.

[81] B. Aksanli, T. Rosing, Providing regulation services and managing data center peak power budgets, in: Design, Automation and Test in Europe Conference and Exhibition, 2014, pp. 1–4, http://dx.doi.org/10.7873/DATE.2014.156.

[82] Z. Liu, A. Wierman, Y. Chen, B. Razon, N. Chen, Data center demand response: avoiding the coincident peak via workload shifting and local generation, in: ACM SIG-METRICS, ISBN: 978-1-4503-1900-3, 2013, pp. 341–342, New York. http://dx.doi.org/10.1145/2465529.2465740.

[83] D.S. Palasamudram, R.K. Sitaraman, B. Urgaonkar, R. Urgaonkar, Using batteries to reduce the power costs of Internet-scale distributed networks, in: 3rd ACM Symposium on Cloud Computing, ISBN: 978-1-4503-1761-0, 2012, pp. 11:1–11:14, New York. http://dx.doi.org/10.1145/2391229.2391240.

[84] J. Novet, Colocation providers, customers trade tips on energy savings, 2013. http://www. datacenterknowledge.com/archives/2013/11/01/colocation-providers-customers-trade-tips-on-energy-savings/.

[85] J. Kerrigan, D. Horowitz, Avison Young Data Center Practice Newsletter, Avison Young, 2014.

[86] Data center colocation market—worldwide market forecast and analysis (2013–2018), 2013. http://www.marketsandmarkets.com/Market-Reports/colocation-market-1252. html.

[87] TeleGeography, Colocation database, https://www.telegeography.com/research-services/colocation-database/.

[88] Data Center Map, Colocation, USA, http://www.datacentermap.com/usa/.

[89] U. Hoelzle, L.A. Barroso, The Datacenter as a Computer: An Introduction to the Design of Warehouse-Scale Machines, first ed., Morgan and Claypool Publishers, San Rafael, CA, 2009, ISBN: 978-1-598-29556-6.

[90] J. Ma, J. Deng, L. Song, Z. Han, Incentive mechanism for demand side management in smart grid using auction, IEEE Trans. Smart Grid 5 (3) (2014) 1379–1388, ISSN 1949-3053, http://dx.doi.org/10.1109/TSG.2014.2302915.

[91] A.-H. Mohsenian-Rad, V.W.S. Wong, J. Jatskevich, R. Schober, A. Leon-Garcia, Autonomous demand-side management based on game-theoretic energy consumption scheduling for the future smart grid, IEEE Trans. Smart Grid 1 (3) (2010) 320–331, ISSN 1949-3053, http://dx.doi.org/10.1109/TSG.2010.2089069.

[92] RAP, Electricity regulation in the US: a guide, Regulatory Assistance Project, 2011.

[93] H. Zhong, L. Xie, Q. Xia, Coupon incentive-based demand response: theory and case study, IEEE Trans. Power Syst. 28 (2) (2013) 1266–1276, ISSN 0885-8950, http://dx. doi.org/10.1109/TPWRS.2012.2218665.

[94] HarborRidge Capital, Colocation data centers: overview, trends & M&A, Harbor Ridge Capital, LLC, 2012. http://www.harborridgecap.com.

[95] D. Wang, C. Ren, A. Sivasubramaniam, B. Urgaonkar, H. Fathy, Energy storage in datacenters: what, where, and how much? SIGMETRICS Perform. Eval. Rev. 40 (1) (2012) 187–198, ISSN 0163-5999, http://dx.doi.org/10.1145/2318857.2254780. http://doi.acm.org/10.1145/2318857.2254780.

[96] S. Martello, P. Toth, Knapsack Problems: Algorithms and Computer Implementations, John Wiley & Sons, Inc., New York, ISBN: 0-471-92420-2, 1990.

ABOUT THE AUTHORS

Thant Zin Oo received the BE degree in electrical systems and electronics at Myanmar Maritime University in 2008. At the same time, he also obtained a BS in Computing and Information System from London Metropolitan University for which he received grant from the British Council. After graduation, he joined Pyae Naing Thu company, Myanmar, where he worked as an electrical engineer. In 2010, he was awarded scholarship for his graduate study at Kyung Hee University, where he is currently working toward Ph.D. degree in computer engineering.

Nguyen H. Tran received the BS degree from Hochiminh City University of Technology and Ph.D. degree from Kyung Hee University, in electrical and computer engineering, in 2005 and 2011, respectively. Since 2012, he has been an assistant professor in the Department of Computer Engineering, Kyung Hee University. His research interest is using queueing theory, optimization theory, control theory, and game theory to design, analyze, and optimize the cutting-edge applications in communication networks, including cloud-computing data center, smart grid, heterogeneous networks, and Internet of Things.

Choong Seon Hong received the BS and MS degrees in electronic engineering from Kyung Hee University, Seoul, Korea, in 1983 and 1985, respectively, and the Ph.D. degree at Keio University in March 1997. In 1988, he joined KT, where he worked on Broadband Networks as a member of the technical staff. In September 1993, he joined Keio University, Japan. He had worked for the Telecommunications Network Lab at KT as a senior member of the technical staff and as a director of the networking research team until August 1999. Since September 1999, he has been working as a professor of the Department of Computer Engineering, Kyung Hee University. He has served as a program committee member and an organizing committee member for International conferences such as SAINT, NOMS, IM, APNOMS, ICOIN, CSNM, ICUIMC, E2EMON, CCNC, ADSN, ICPP, DIM, WISA, BcN, ManFI, TINA, etc. His research interests include future Internet, wireless networks, network security, and network management. He is a senior member of the IEEE and a member of the ACM, IEICE, IPSJ, KICS, KIISE, KIPS, and OSIA.

Shaolei Ren received his B.E., M.Phil. and Ph.D. degrees, all in electrical engineering, from Tsinghua University in 2006, Hong Kong University of Science and Technology in 2008, and University of California, Los Angeles, in 2012, respectively. From 2012 to 2015, he was with Florida International University as an Assistant Professor. Since July 2015, he has been an Assistant Professor at University of California, Riverside. His research interests include power-aware computing, data center resource management, and network economics.

Gang Quan received his Ph.D. from the Department of Computer Science & Engineering, University of Notre Dame, USA, his M.S. from the Chinese Academy of Sciences, China, and his B.S. from the Department of Electronic Engineering, Tsinghua University, Beijing, China. He is currently an associate professor at the Department of Electrical and Computing Engineering, Florida International University. Before he joined the department, he was an assistant professor at the Department of Computer Science and Engineering, University of South Carolina. His research interests include real-time systems, embedded system design, power-/thermal-aware computing, advanced computer architecture, reconfigurable computing.

CHAPTER TWO

Energy-Efficient Big Data Analytics in Datacenters

Farhad Mehdipour*, Hamid Noori[†], Bahman Javadi[‡]
*E-JUST Center, Graduate School of Information Science and Electronics Engineering, Kyushu University, Fukuoka, Japan
[†]Engineering Department, Ferdowsi University of Mashhad, Mashhad, Iran
[‡]School of Computing, Engineering and Mathematics, University of Western Sydney, Sydney, New South Wales, Australia

Contents

Abstract

The volume of generated data increases by the rapid growth of Internet of Things, leading to the big data proliferation and more opportunities for datacenters. Highly virtualized cloud-based datacenters are currently considered for big data analytics. However, big data requires datacenters with promoted infrastructure capable of undertaking more responsibilities for handling and analyzing data. Also, as the scale of the datacenter is increasingly expanding, minimizing energy consumption and operational cost is a vital concern. Future datacenters infrastructure including interconnection network, storage, and servers should be able to handle big data applications in an energy-efficient way. In this chapter, we aim to explore different aspects of could-based datacenters for big data analytics. First, the datacenter architecture including computing and networking technologies as well as datacenters for cloud-based services will be illustrated. Then the concept of big data, cloud computing, and some of the existing cloud-based datacenter platforms including tools for big data analytics will be introduced. We later discuss the techniques for improving energy efficiency in the cloud-based datacenters for big data analytics. Finally, the current and future trends for datacenters in particular with respect to energy consumption to support big data analytics will be discussed.

ABBREVIATIONS

AIS all-in strategy
APU accelerated processing unit
BPaaS business process as-a-service
CEP complex event processing
CPU central processing unit
CS covering set
DAaaS data analytic as-a-service
DAG directed acyclic graph
DBMS database management system
DCN datacenter network
DFS distributed file system
DVFS dynamic voltage and frequency scaling
E/O electrical-to-optical
FPGA field-programmable gate array
Gbps gigabits per second
GFS Google file system
GPU graphics processing unit
HDFS Hadoop distributed file system
HPC high-performance computing
HVAC heating, ventilation, and air conditioning
IaaS infrastructure as-a-service
IO (I/O) input/output
IoT Internet of Things
IT information technology
KaaS knowledge as-a-service
kWh kilo-Watt hour
MPI message passing interface
MtCO$_2$e metric tons of carbon dioxide equivalent
MW mega Watt
NoSQL not only SQL
O/E optical-to-electrical
OFDM orthogonal frequency-division multiplexing

OS operating system
PaaS platform as-a-service
PB petabytes
PB/s petabytes per second
PFLOPS peta floating-point operations per second
QoS quality of service
RDBMS relational database management system
SaaS software as-a-service
SAN storage area network
SIMD single instruction multiple data
SQL structured query language
Tbps terabits per second
TCP transmission control protocol
ToR top-of-the-rack
UPS uninterruptible power supply
VM virtual machine
WDM wavelength division multiplexing

1. INTRODUCTION

Due to the latest advances in information technology, the volume of generated data further increases by the rapid growth of cloud computing and the Internet of Things (IoT). Widely distributed sensors in an IoT collect and transmit data that should be integrated, analyzed, and stored. Such data referred as "big data," far surpasses the available computing capacity in quantity, heterogeneity, and speed. The emergence of big data brings excellent development opportunities to datacenters that are rapidly evolving. There is also a great attention to cloud services, virtualization solutions, and high-performance computing (HPC) to boost service velocity and business agility, support big data analytics, and improve datacenter economics. Conventional computing systems are giving way to highly virtualized environments that reshape datacenter traffic flows and dramatically affect datacenter network (DCN) designs [1].

Big data requires promotion of datacenter infrastructure in both hardware and software aspects. With the continued growth of the volumes of structured and unstructured data, the data processing and computing capacities of the datacenter shall be greatly enhanced. Although current datacenters provide hardware facilities and storage for data, it is crucial for them to undertake more responsibilities, such as acquiring, analyzing, and organizing the data [2]. Also, as the scale of the datacenter is increasingly expanding, it is also an important issue on how to reduce the operational cost for the development of datacenters.

Power consumption is a major concern in the design and development of modern datacenters. Although the performance per watt ratio has been constantly rising, the total power consumed by computing systems is hardly decreasing [3]. It has been shown that in 2007, the worldwide demand for electricity was around 330 Billion kW h from datacenters [4]. This amount is similar to the electricity consumption in the whole UK. Given the demand for more datacenters and emergent of complex big data processing, it is estimated that the electricity consumption will be triple by 2020 (more than 1000 Billion kW h). This also has a major impact on the environment by greenhouse gas emissions. By 2020, the carbon footprint will be about 275 metric tons of carbon dioxide equivalent ($MtCO_2e$) that is more than the double of what we had in 2007 [5]. Therefore, novel power-efficient technologies must be developed to minimize the power consumption in datacenters, especially to handle big data analytics.

In this chapter, we aim to explore different aspects of could-based datacenters for big data analytics. In Section 2, datacenter architecture including computing and networking technologies as well as datacenters for cloud-based services will be described. In Section 3, the concept of big data, cloud computing some of the existing cloud-based datacenter platforms and tools for big data analytics will be introduced. Section 4 discusses the techniques for improving energy efficiency in the cloud-based datacenters for big data analytic. Section 5 highlights the current and future trends for datacenters in particular with respect to energy consumption to support big data analytics. Section 6 summarizes this chapter.

2. DATACENTER AND CLOUD COMPUTING

A datacenter is usually known as the infrastructure used by enterprises and is designed to host computing and networking systems and components for IT service demands. A datacenter typically involves storing, processing, and serving large amounts of mission-critical data to clients in client/server architecture. Redundant or backup power supplies, cooling systems, redundant networking connections, and security systems for running the enterprise's core applications are extensively needed in datacenters. Datacenter management requires the high reliability of both the connections to the datacenter as well as the information stored within the datacenter's storage.

Scheduling of application workloads on the available compute resources cost-effectively is another issue that should be addressed in datacenters. Datacenter size in terms of energy consumption varies widely, ranging from 1 MW to more than 30 MW [6].

2.1 Datacenter Architecture

The design of a datacenter is often divided into four categories as "Tier I–IV." The four-tier classification loosely based on the power distribution, uninterruptible power supply (UPS), cooling delivery, and redundancy of the datacenter. Tier I datacenters have a single path for power distribution, UPS, and cooling distribution, without redundant components, while Tier II adds redundant components to enhance availability. Tier III datacenters have one active and one alternate distribution path for utilities that provide redundancy even during maintenance. Tier IV datacenters have two simultaneously active power and cooling distribution paths, redundant components in each path and are supposed to tolerate any single equipment failure without impacting the load [6].

The high-level block diagram of a typical datacenter is shown in Fig. 1. A datacenter consists of multiple racks hosting the servers connected through a top-of-the-rack (ToR) switch which are further interconnected through aggregate switches in a tree topology. The DCN interconnects a massive number of servers and provides routing service for the data flowing through the network among computing elements. The tree architecture is a well-known and common interconnection scheme in the current datacenters. However, it suffers from the problems of low scalability, high cost as well as a single point of failure. An interconnection network should be modular with a fast reconfiguration capability, and a high capacity and scalability [8,9]. In the current datacenters, the network is usually a canonical fat-tree Tier II or Tier III architecture (Fig. 1) [7,10]. In the Tier III topologies (shown in the figure), one more level is applied in which the aggregate switches are connected in a fat-tree topology using the core switches. When a user issues a request, a packet is forwarded through the Internet to the front end of the datacenter. In the front end, the content switches and the load balance devices are used to route the request to the appropriate server. Most of the current datacenters are based on commodity switches for the interconnection network. The topology of the network interconnecting the servers significantly affects the agility and reconfigurability of the datacenter infrastructure to respond to changing application demands and service requirements [100].

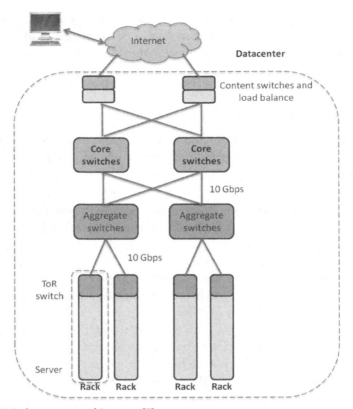

Figure 1 A datacenter architecture [7].

There are alternate approaches to scalable, cost-effective network architectures which can be grouped into fixed topology and flexible topology
networks. Fixed topology networks can be further categorized into two
topologies: tree-based topologies such as fat-tree [3,100] and Clos Network
[12], and recursive topologies such as DCell [13] and BCube [14]. Flexible
topologies have the ability to adapt their network topology based on the traffic demand, at run time. Examples of such topologies include c–Through
[15], Helios [8], and OSA [16]. Every DCN has its unique approaches
for network topology, routing algorithms, fault tolerance, and fault recovery
[100].

Traditional datacenter architectures including networking and storage
resources support the needs of specific client–server applications. The capability of existing datacenter for data processing is limited to compute and
storage infrastructures within a local area network, e.g., a single cluster

within a datacenter. Further, the conventional datacenters built for client–server computing lack in essential capabilities to meet the requirements of today's highly virtualized compute environments; they cannot meet the performance and availability demands; make inefficient use of network resources; do not scale in a linear fashion; and are not well suited for the high-bandwidth, low-latency server-to-server traffic flows that is common in current datacenters [1].

However, advances such as powerful multicore processor-based servers, virtualization, cloud, and distributed computing are transforming the datacenters to incorporate new features. Today's datacenters contain thousands of switches and servers, run data-intensive applications from cloud services such as search, Web e-mail, to infrastructural computations such as MapReduce [17]. Many IT organizations are consolidating compute, network, and storage resources and employing virtualization solutions and cloud-based services based on new networking models.

2.2 Cloud Computing

Cloud computing is the integration of computing and data infrastructures to provide a scalable, agile, and cost-effective approach to supporting the ever-growing critical IT needs [9]. Cloud computing relieves its users from the burdens of provisioning and managing their datacenters and allows them to pay for resources only when needed (i.e., pay-as-you-go). It provides more profit to both service providers who capitalize poorly utilized resources and users who only need to pay per their use. Furthermore, increasing the resource utilization results in less energy consumption and carbon footprint that can increase the economical profit of the cloud for providers [3].

2.2.1 Virtualization in Cloud Systems

Physical resources can be divided into a number of logical slices called virtual machines (VMs). Each VM can accommodate an individual operating system (OS) creating for the user a view of a dedicated physical resource and ensuring the performance and failure isolation between VMs that are sharing a single physical machine. The virtualization layer lies between the hardware and OS and, therefore, a virtual machine monitor takes the control over resource sharing/multiplexing and has to be involved in the system's power management [3]. The virtualization technology provides the ability to encapsulate the workload in VMs and consolidate them to a single physical server. The consolidation has become especially effective after the adoption of multicore central processing units (CPUs) in computing environments, as many VMs can be allocated to a single physical node leading to the improved utilization of resources and reduced energy consumption compared to a multinode setup [3,18].

2.3 Datacenters for Cloud Computing Services

Datacenter infrastructure has been receiving significant interest due to their highlighted role in supporting the rapidly growing cloud-based applications. Datacenters including the computing, storage, and communication resources form the core of the support infrastructures for cloud computing. With the proliferation of cloud computing and cloud-based services, datacenters are becoming increasingly large with massive number of servers and a huge amount of storage. Server and even network virtualization are increasingly employed to make the datacenter flexible and adapt to varying demands [9].

2.3.1 Multicloud Systems

Datacenters based on the interconnection of multiple cloud systems provide an opportunity to improve overall quality of services (QoSs) parameters by deploying and executing applications over a large cluster of resources [19]. Such datacenters are heterogeneous, distributed, and highly uncertain. The system becomes more complicated where mobile devices are considered as cloud nodes as well. The availability, performance, and state of resources, applications, and data may continuously change in a multicloud datacenter. Uncertainty is a fact in multicloud datacenter environments due to several reasons such as security attacks, network and resource failures, incomplete global knowledge among competing applications leading to nonoptimized use of the resources and network bandwidth, etc. The current data–intensive or big data application programming frameworks such as MapReduce and workflow models are inadequate to handle these issues [20].

3. HANDLING BIG DATA ON CLOUD-BASED DATACENTERS

Nowadays, the data have reached in different directions in terms of size, type, and speed and have received wide attention as "big data." The huge amount of data coming from various sources with different types and in high speeds needs appropriate infrastructure for efficient processing and knowledge discovery. Currently, datacenters incorporate new features such as virtualized computing in the cloud systems that make them a candidate platform for processing the big data.

3.1 Big Data

Big data refers to the large amounts (volume) of heterogeneous data (variety) that flows continuously (velocity) within data-centric applications. We can classify big data requirements based on its five main characteristics [21]:

- *Volume*: This is the primary characteristic of big data, refers to the large size (tera or petabytes) of records, transactions, tables, files, video, Web text, and sensor logs, etc. As the size of data to be processed is large, it needs to be broken into manageable chunks. Data needs to be processed simultaneously across multiple systems and several program modules.
- *Variety*: The big data comes from a great variety of sources. Data of different formats, types, and structures need to be processed using efficient solutions.
- *Velocity*: It refers to the frequency of data generation or the frequency of data delivery. The stream of data acquired via multiple sources needs to be processed at a reasonable speed.
- *Ambiguity*: Big data is ambiguous by nature due to the lack of relevant metadata and context in many cases. For example, "M" and "F" in a data set can mean, Monday and Friday, male and female, or mother and father, respectively.
- *Complexity*: Big data complexity needs to use many algorithms to process data quickly and efficiently. Several types of data need multistep processing, and the scalability is extremely important as well. Processing large-scale data requires an extremely HPC environment that can be easily managed and its performance can be tuned with linear scalability.

Big data analytics describes the process of performing complex analytical tasks on data that typically includes grouping, aggregation, or iterative processes. The increase in the amount of data raises several issues for analysis software: (1) The amount of data is increasing continuously at a high speed, yet data should be up-to-date for analysis tasks, (2) The response time of a query grows with the amount of data.

At the same time, latencies must be reduced to provide actionable intelligence at the right time, (3) analysis tasks need to produce query results on large data sets in an adequate amount of time [22].

Figure 2 shows a typical flow for the big data processing. The first step after the acquisition of data is to perform fusion/integration of the data

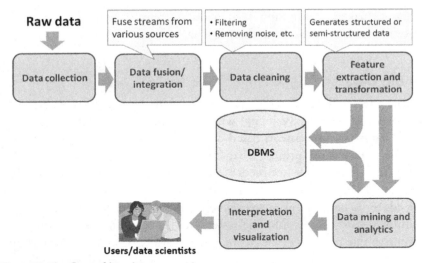

Figure 2 The flow of big data processing.

coming from multiple sources. Data cleaning is the next step that may consume large processing time though it may significantly reduce the data size that leads to less time and effort needed for data analytics. The raw data are normally unstructured that neither have a predefined data model nor are organized in a predefined manner. Thus, the data are transformed to semi-structured or structured data in the next step of the flow [23].

The main phase of big data processing is to perform discovery of data, which is where the complexity of processing data lies. A unique characteristic of big data is the manner in which the value is discovered. It differs from conventional business intelligence, where the simple summing of known values generates a result. The data analytics is performed through visualizations, interactive knowledge-based queries, or machine learning algorithms that can discover knowledge [4]. Due to the variant nature of the data, there may not be a single solution for the data analytics problem so, the algorithm may be short-lived.

3.2 Big Data and Cloud Computing

As aforementioned, cloud computing allows access to information and computer resources from anywhere that a network connection is available. It supplies large networks of virtualized services: hardware resources (CPU, storage, and network) and software resources (e.g., databases, load balancers). The datacenter cloud facilitates virtual centralization of application, computing, and data. Nowadays, shifting the big data applications from the physical infrastructure into the virtualized datacenters in computational clouds is a major trend [20]. While cloud computing optimizes the use of resources, it does not provide an effective solution hosting big data applications yet. Gartner's Inc. Hype Cycle states that cloud computing and big data are the fastest growing technologies that dominate the research world for the next decade [24].

Figure 3 shows the high-level components that are building blocks of the next-generation cloud-based datacenters supporting big data [23,25]. The lowest layer represents various forms of the data generated by different sources. The next layer interfaces data stream moving toward datacenter cloud system. It may involve real-time processing of data and technologies such as complex event processing (CEP [101]) as well. The next tier represents the technologies that will be used in integrating, cleaning, and transforming the data to multiple types and sources of data. Storage and distributed data management technologies such as NoSQL for unstructured and semi-structured data and the relational database management system (RDBMS) for structured data are represented in the next tier. The topmost layer indicates the analytical layer that will be used to drive the needs of the big data analytics.

The cloud computing system comprises three service models: infrastructure as-a-service (IaaS), platform as-a-service (PaaS), and software as-a-service (SaaS). A classification of big data functionalities against cloud service models is illustrated in Fig. 4 [4]. At the lowest level, current hardware infrastructure is virtualized with cloud technologies, and the hardware

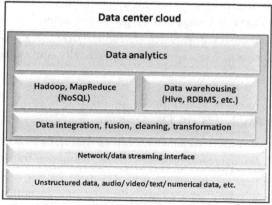

Figure 3 Components of next-generation cloud-based datacenter.

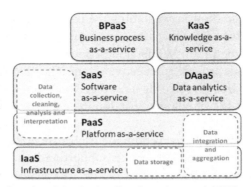

Figure 4 Big data functionalities in the Cloud service model [27].

infrastructure, as well as platforms, will be provided as services. In the upper layer SaaS and, on top of it, business processes as a service can be built. Big data can be defined as a *data analytic as-a-service* (DAaaS), an additional cloud layer located between PaaS and SaaS layers. Some of the data analytics functionalities can be implemented through existing tools in SaaS and can be used with minor customizations, or through the software developed in PaaS. Furthermore, most of the cloud services offering big data capabilities can be classified as IaaS, with minor exceptions. It is because different platforms tend to cover all aspects of big data from data capturing to storage and analysis, and they include the underlying infrastructure needed to handle all the required capabilities, particularly those of the storage.

Despite rather remarkable number of cloud-based big data solutions that have been released on the market (e.g., OpenCrowd [28], Infochimps Platform [29], Opani [28], PiCloud [102], Oracle big data appliance, and IBM Infosphere BigInsights [21]), many of them fall shortly of big data challenge. It is therefore required to provide computing infrastructures such as specialized datacenters, software frameworks, data processing technologies, and distributed storage system specifically optimized for big data applications.

3.3 Big Data Analytics on Datacenters

The major driver for modern datacenters is analyzing of large-scale and massive data sets [6]. So, datacenters need to have adequate infrastructure such as interconnection networks, storage, and servers to handle big data applications. While these resources are currently used in datacenters, they should be evaluated for handling a new type of data-centric applications. In the following, these resources including power consumption as a main concern for future datacenters will be reviewed.

3.3.1 Interconnection Networks

The interconnection network in datacenters is the core for supporting big data and is the most important infrastructure that needs urgent attention. Big data applications transfer large-scale data to datacenters for processing and analysis. So, we need to consider two types of interconnection: inter-datacenter connection and intra-datacenter connection.

(i) *Inter-datacenter connection*: This is the connection from the outside world to the datacenter, which is normally based on the existing network infrastructure (i.e., Internet). The modern physical infrastructure for most of the countries is the high-bandwidth fiber optic interconnections that can handle the rapid growth of data size. The common architecture in these interconnections is IP-based wavelength division multiplexing (WDM), which is based on multiple optical carriers multiplexing in different wavelength [8,98]. In this technology, each optical fiber can carry several signals with a different wavelength and, therefore, increase the network bandwidth. While most of the WDM-based networks have deployed with the bandwidth of 40 Gbps, recently a new standard for 100 Gbps is introduced to address the demand for the high-speed inter-datacenter connection [31].

Since WDM technology is limited by the bandwidth of the electronic bottleneck, a new technology called orthogonal frequency-division multiplexing (OFDM) is introduced. OFDM is a multicarrier parallel transmission technology, which divides the high-speed data flow into low-speed sub-data flow to transmit them over multiple orthogonal subcarriers [32]. This makes OFDM more flexible and efficient compared with the WDM technology and more promising for inter-datacenter connection that is dealing with big data transferring.

(ii) *Intra-datacenter connection*: Intra-datacenter connection is used to transfer the data within a datacenter. As mentioned earlier, most of the current datacenters consist of high-performance servers interconnected with commodity switches in a multilevel fat-tree topology [6]. While servers in each rack are connected through 1 Gbps links, all the rack servers are interconnected through a set of 10 Gbps switches. Finally, these switches will be connected through a set of core switches with 100 Gbps links (using a bundle of 10 Gbps links). Although this architecture is scalable and fault tolerant, power consumption is the main issue. The other issue in this architecture is the high-latency due to multiple store-and-forward in the switches hierarchy. For big data applications, we need a network technology for intra-datacenter connection, which can provide high throughput, low latency, and low-energy consumption.

To solve these issues, *optical networks* recently have received great attention as an alternative for interconnection networks in datacenters to address the demands of big data analytics. Currently, optical networks are only used for point-to-point communications in datacenters where the links are based on low-cost multimode fibers for short-reach communications [7]. The desired future architecture for intra-datacenter interconnection will be all-optical connections with switching in the optical domain as shown in Fig. 5. In this technology, electrical-to-optical (E/O) and optical-to-electrical (O/E) transceivers will be completely removed, which are the main source of power consumption in the current architecture. Therefore, all-optical interconnection networks can provide bandwidth in terabits per second (Tbps) scale with low-power consumption for future DCNs.

3.3.2 Storage

An efficient storage mechanism for big data is an essential part of the modern datacenters. The main requirement for big data storage is file systems that is the foundation for applications in higher levels. The Google file system

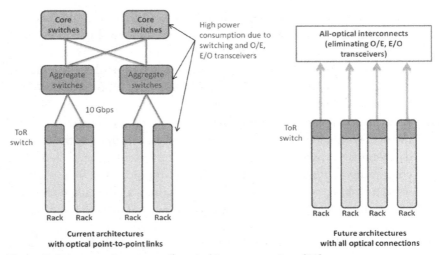

Figure 5 Point-to-point versus all-optical interconnections [33].

(GFS) is a distributed file system (DFS) for data-centric applications with robustness, scalability, and reliability [8]. GFS can be implemented in commodity servers to support large-scale file applications with high performance and high reliability. Colossus is the next generation of GFS with better reliability that can handle small files with higher performance [34].

Hadoop distributed file system (HDFS) is another file system used by MapReduce model where data are placed more closely to where they are processed [35]. HDFS uses partitioning and replication to increase the fault tolerance and performance of large-scale data set processing. Another file system for storing a large amount of data is Haystack [36] which is used by Facebook for handling a lot of photos storing in this Web site. This storage system has a very low overhead that minimizes the image retrieval time for users. The above-mentioned file systems are the results of many years research and practice so can be utilized for big data storage.

The second component in big data storage is a database management system (DBMS). Although database technology has been advancing for more than 30 years, they are not able to meet the requirements for big data. Nontraditional relational databases (NoSQL) are a possible solution for big data storage, which are widely used recently. In the following, we review the existing database solutions for big data storage in three categories: key-value databases, column-oriented databases, and document-oriented databases.

Key-values databases normally have simple data model and data are stored based on key-values. So they have a simple structure with high expandability and performance compared to relational databases. *Column-oriented databases* use columns instead of the row to process and store data. In these databases, both columns and rows will be distributed across multiple nodes to increase expandability.

Document-oriented databases are designed to handle more complex data forms. Since documents can be in any modes (i.e., semi-structured data), so there is no need for mode migration. Table 1 shows the list of big data storages that are classified into three types. These databases are available to handle big data in datacenters and cloud computing systems.

3.3.3 Processing and Analysis Tools and Techniques

Big data processing is a set of techniques or programming models to access large-scale data to extract useful information for supporting and providing decisions. In the following, we review some tools and techniques, which are available for big data analysis in datacenters.

Table 1 List of Databases for Big Data Storage in Datacenters

Database	Type	Description
Amazon: Dynamo [20]	Key–value databases	A distributed, high reliable, and scalable database systems used by Amazon for internal applications
Voldemort [37]	Key–value databases	A storage system used in LinkedIn Web site
Google: Bigtable [38]	Column-oriented databases	A distributed storage system used by several Google produces, which as Google Docs, Google Maps, and Google search engine. Bigtable can handle data storage in the scale of petabytes using thousands of servers
Cassandra [9]	Column-oriented databases	A storage system developed by Facebook to store large-scale structured data across multiple commodity servers. Cassandra is a decentralized database that provide high availability, scalability, and fault tolerance
Amazon Simple DB [39]	Document-oriented databases	A distributed database designed for structured data storage and provided by Amazon as the Web service
CouchDB [27]	Document-oriented databases	Apache CouchDB is document-based storage system where JavaScript is used to query and manipulate the documents

As mentioned in previous section, big data usually stored in thousands of commodity servers so traditional programming models such as message passing interface (MPI) [40] cannot handle them effectively. Therefore, new parallel programming models are utilized to improve the performance of NoSQL databases in datacenters. MapReduce [17] is one of the most popular programming models for big data processing using large-scale commodity clusters. MapReduce is proposed by Google and developed by Yahoo. Map and Reduce functions are programmed by users to process the big data distributed across multiple heterogeneous nodes. The main advantage of this programming model is simplicity, so users can easily utilize that for big data processing. A certain set of wrappers is being developed for MapReduce. These wrappers can provide a better control over the MapReduce code and aid in the source code development. Apache Pig is a structured query language (SQL)-like environment developed at Yahoo [41] is being used by many organizations like Yahoo, Twitter, AOL, LinkedIn, etc. Hive is another MapReduce wrapper developed by Facebook [42]. These two wrappers provide a better environment and make the code development simpler since the programmers do not have to deal with the complexities of MapReduce coding.

Hadoop [43,44] is the open-source implementation of MapReduce and is widely used for big data processing. This software is even available through some Cloud providers such as Amazon EMR [96] to create Hadoop clusters to process big data using Amazon EC2 resources [45]. Hadoop adopts the HDFS file system, which is explained in previous section. By using this file system, data will be located close to the processing node to minimize the communication overhead. Windows Azure also uses a MapReduce runtime called Daytona [46], which utilized Azure's Cloud infrastructure as the scalable storage system for data processing.

There are several new implementations of Hadoop to overcome its performance issues such as slowness to load data and the lack of reuse of data [47,48]. For instance, Starfish [47] is a Hadoop-based framework, which aimed to improve the performance of MapReduce jobs using data lifecycle in analytics. It also uses job profiling and workflow optimization to reduce the impact of unbalance data during the job execution. Starfish is a self-tuning system based on user requirements and system workloads without any need from users to configure or change the settings or parameters. Moreover, Starfish's Elastisizer can automate the decision making for creating optimized Hadoop clusters using a mix of simulation and model-based estimation to find the best answers for what-if questions about workload performance.

Spark [49], developed at the University of California at Berkeley, is an alternative to Hadoop, which is designed to overcome the disk I/O limitations and improve the performance of earlier systems. The major feature of Spark that makes it unique is its ability to perform in-memory computations. It allows the data to be cached in memory, thus eliminating the Hadoop's disk overhead limitation for iterative tasks. The Spark developers have also proposed an entire data processing stack called Berkeley data analytics stack [50].

Similarly, there are other proposed techniques for profiling of MapReduce applications to find possible bottlenecks and simulate various scenarios for performance analysis of the modified applications [48]. This trend reveals that using simple Hadoop setup would not be efficient for big data analytics, and new tools and techniques to automate provisioning decisions should be designed and developed. This possibly can be a new service (i.e., *big data analytics as-a-service*) that should be provided by the Cloud providers for automatic big data analytics on datacenters.

In addition to MapReduce, there are other existing programming models that can be used for big data processing in datacenters such as Dryad [51] and Pregel [52]. Dryad is a distributed execution engine to run big data applications in the form of directed acyclic graph (DAG). Operation in the vertexes will be run in clusters where data will be transferred using data channels including documents, transmission control protocol (TCP) connections, and shared memory. Moreover, any type of data can be directly transferred between nodes. While MapReduce only support single input and output set, users can use any number of input and output data in Dryad. Pregel is used by Google to process large-scale graphs for various purposes such as analysis of network graphs and social networking services. Applications are introduced as directed graphs to Pregel where each vertex is modifiable, and user–defined value and edge show the source and destination vertexes.

3.3.4 Platforms for Big Data Analytics
Solutions for the issue of the growing computation power required by big data analytics fall in two different categories. On approach is to scale the current systems and the other one is to look for more suitable and efficient systems as a substitution for current servers. Scaling is the ability of the system to adapt to increased demands for big data processing. Horizontal scaling and vertical scaling are two general approaches.

(i) *Vertical scaling* (also known as scale up) includes empowering machines with more memory and higher performance processors as well as involving specialized hardware such as accelerators. The most popular vertical scale up paradigms are multicore processors, graphics processing unit (GPU), accelerated processing units (APUs), field-programmable gate array (FPGA), and HPC clusters. HPC clusters [53], also called as blades or supercomputers, are machines with thousands of cores. They can have a different variety of disk organization, cache, communication mechanism, etc., depending upon the user requirement. MPI is typically used as the communication scheme for such platforms [40]. Since there are software and programming tools that for easy of vertical scaling (e.g., having multicore processors, GPUs), most of the software can take advantage of it transparently. However, vertical scaling needs substantial financial investments at one point in time. Furthermore, to cope with future workloads, the system needs to be adequately powerful, and initially, the additional performance goes to waste.

(ii) *Horizontal scaling* includes extending clusters by adding more servers and distributing the work across many servers, which are usually commodity machines. This increases performance in smaller steps, and the financial investment to upgrade is by far smaller. However, to use these large clusters including many servers efficiently, an analysis software (parallel programming utilities) on top of them has to be developed to handle distribution by itself. Parallel programming uses a divide and conquer approach to breaking down the application into independently processable partitions and send for distributed machines for processing. After finishing the process, the intermediate results have to be merged into a final result. Synchronization among the parallel processors is one of the essentials [22]. Different approaches have evolved for horizontal scaling over the time: parallel systems using an SQL-like interface and massive data analysis systems. Some of the prominent horizontal scale-out platforms include peer-to-peer networks [37,39], parallel database systems, Hyracks [54], and Stratosphere [55], since they are samples of infrastructures and tools that support horizontal scaling and provide the facilities to use many machines available in a cluster.

Peer-to-peer networks [37,39] are decentralized and distributed network architecture including millions of connected machines. Each node (machine) in the network serves as well as consumes resources. MPI is the typical communication approach for exchanging the data among nodes. In this structure, scaling out can be unlimited (in case of support of software). The major challenge of these networks is in the communication among different nodes. In case of initiating broadcasting messages in such a network, the messages are sent from a root that is an arbitrary in the form of a spanning tree over the network. The broadcasting messages are cheaper than the aggregation of data/results that are much more expensive [40].

Parallel DBMSs such as GRACE [97], Gamma [57], and Teradata [58] are pioneers of parallel relational database systems. Classical DBMSs use an n-ary storage model to provide intra record locality. A relation consists of tuples with a defined set of attributes. Gamma [57] is a parallel relational database machine that exploits data-flow query processing techniques. Each operator in a relational query tree is executed by one or more processes. The scheduler places these processes on a combination of processors with and without disk drives. In Gamma's hardware design, associated with each disk drive there is a processor, and the processors are interconnected via an interconnection network. This design provides high disk bandwidth and permits the I/O bandwidth to be expanded incrementally. GRACE [97] is a parallel relational database system adopting data stream-oriented processing. Teradata [58] is a commercial system that has built a DBMS for decision support using a parallel architecture and multiple processors.

Hyracks [54] is a data parallel platform for data-intensive computations that has been developed by the computer science group of the University of California, Irvine. It features a parallel data-flow execution model. Jobs are specified as DAGs composed of operators, represented as nodes, and connectors, represented as edges. Hyracks abstracts the MapReduce data model and uses tuples with fields of arbitrary types.

Stratosphere [55] system has been built by a research team comprised of TU Berlin, HU Berlin, and HPI Potsdam since 2008, consists of the PACT programming model and the massively parallel execution engine Nephele. The PACT programming model is a generalization of the MapReduce programming model and operates on a key/value data model.

On the other hand, according to some of current studies, current servers do not match well with the computational requirement of big data analytics. Therefore, they suggest looking for more efficient hardware platforms for future servers such as Microservers [31,59–63], scale-out processors [27,56,64,65], and system-level integration [11].

4. ENERGY EFFICIENCY IN DATACENTERS FOR BIG DATA ANALYTICS

Today, datacenters and servers consume enormous amounts of energy and, therefore, improving energy consumption in datacenters is crucial. In a typical datacenter, servers, storage, and network devices consume around 40%, 37%, and 23% of the total IT power, respectively [12]. Handling big data in datacenters has strong impacts on the power consumption of all three components. In addition, increase of power consumption in datacenters raises the power usage of the HVAC equipment (heating, ventilation, and air conditioning) to keep the whole system at the working temperature. Therefore, by improving the power efficiency of any component in datacenters, the overall power consumption decreases significantly due to indirect power saving in the HVAC equipment. Reducing power consumption in hardware and network architecture [3,34,66], and smart cooling technologies [26] are effective methods to save energy in datacenters. It has been shown that saving 1.0 W power in the IT devices will save around 2.5 W in the total power [35]. These evidence indicate that we are in an urgent need for new technologies and techniques to improve the power consumption of datacenters especially in the presence of big data applications.

In a highly connected datacenter, in practice only a few devices are working at a given time. Therefore, a significant amount of energy can be saved if we can power off those idle devices. Intel Research proposed a proxy architecture that uses a minimal set of servers to support different forms of idle-time behavior for saving energy [30]. In such case, the routing can be adjusted so that data traffic flow through a certain set of devices.

As the network interconnect infrastructure is a significant energy consumer in datacenters, therefore, it is essential to develop intelligent techniques to manage network resources efficiently. Further, optical interconnection networks provide a solution for scalable architectures with optimized energy efficiency [67].

In the virtualized datacenters, one possible way of reducing energy consumption is to optimize network topologies established between VMs and thus reduce the network communication overhead and the load of network devices steadily. A reduction in the transition overhead caused by switching between different power states and the VM migration overhead can also greatly advance the energy-efficient resource management. Rizvandi et al. [24] propose consolidation of applications for significant reduction in energy consumption of cloud-based systems. As VMs are increasingly populating the datacenter (42% more per year [33]), by migrating VMs to proper servers the number of active and operating servers can be reduced [5].

In the Cloud federations comprising geographically distributed datacenters, efficient distribution of the workload across geographically distributed datacenters can reduce the costs by dynamically reallocating the workload to a place where the computing resources, energy, and/or cooling are cheaper (e.g., solar energy during daytime across different time zones). Moreover, a power-aware OS can monitor the overall system's performance and appropriately apply power management techniques (e.g., dynamic voltage and frequency scaling, DVFS) to the system components [3].

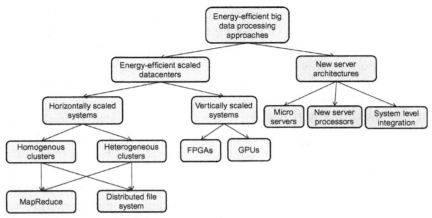

Figure 6 Classification of various approaches for energy-efficient big data processing.

Regarding the low-level system design, it is important to improve the efficiency of power supplies and develop hardware components supporting the performance scaling proportionally to the power consumption. It is also necessary to develop energy-efficient resource management approaches that leverage multicore CPUs due to their wide adoption. In this section, we discuss some energy-efficient approaches for big data processing. Figure 6 shows the general approaches for energy-efficient big data processing. The approaches can be broadly divided into two groups. One group focuses on energy-aware scaling of datacenters including both vertically and horizontally. The second group believes in a new architecture for servers as a replacement for currently used machines for energy enhancement. This is because, according to some studies the architectures of current servers does not match well with the computational requirements of big data processing applications. Using low-power processors (microservers), more system-level integration and a new architecture for servers' processors are some of the solutions that have been discussed recently as an energy-efficient replacement for current machines.

4.1 Energy-Efficient Scaled Datacenters

In this section, we discuss some energy-efficient scaling approaches including both horizontally and vertically. In these approaches, while the system is scaled to support computational requirements for big data processing, the efficiency of the system in term of energy is taken into account as well. The energy-efficient horizontally scaled approaches mainly focus on

MapReduce programming model and scheduling as well as DFS or both. In vertical scaling, FPGAs and GPUs are considered as accelerators for enhancing energy efficiency.

4.1.1 Horizontally Scaled Systems

MapReduce is used widely by many industries for big data processing through spreading data across datacenters or a large number of clusters. Thus, power management for MapReduce clusters has also become important [44,68]. Unfortunately, the innate features of conventional MapReduce programming model result in energy inefficiency. Examples of these features include [44]

- replicating data across multiple machines for fault tolerance,
- low-average utilization of machines in large datacenters, due to variable distributions of the jobs arrival have,
- sharing MapReduce clusters to run jobs with different behavior (some being CPU-intensive, and some I/O-intensive), and
- speculative tasks execution performed by MapReduce to ensure reliability.

Combining workloads on a fewer servers and powering down the idle servers is a common approach to enhance the energy efficiency. However, considering the characteristics of MapReduce, mentioned before, such techniques will not work so effectively because they [44]

i. affect parallelism, and MapReduce performance is proportional to the number of machines due to its parallel nature,

ii. increase latencies because data would have to be fetched from remote locations,

iii. cause data unavailability as data are distributed across servers, and

iv. increase network traffic due to data replication done by NameNode [69].

Therefore, effective energy management techniques should be aware of the underlying DFS and programming frameworks otherwise the application performance degrades significantly [44].

Most of the techniques that enhance the energy efficiency in MapReduce environments attempt to reduce the idle periods on nodes with a less number of active nodes in the cluster. To do so, they modify the MapReduce programming model and DFS to effectively and intelligently (1) merge the jobs, (2) redistribute the data, and (3) reconfigure the nodes [44]. Tiwari [44] classifies the energy-efficient MapReduce techniques into two MapReduced-based and HDFS-based techniques. Then, he categorizes the MapReduce programming model modifications for consolidating/distributing workload based on (1) workload characteristics and (2) hardware characteristics [44]. Also, HDFS cluster modification techniques modify HDFS component work by consolidating the data on fewer active nodes so that other nodes can go to sleep state. According to Tiwari [44], HDFS cluster modification techniques are classified as (1) Replica binning-based data placement and zoning strategies and (2) Temporal binning-based data placement and zoning strategies. The former techniques ensure that at least one copy of every data block is always available. The latter techniques consider the data access patterns based on time and workload to create the zones so that frequently accessed data are always available. Wirtz [68] proposes another classification for energy efficiency techniques for MapReduce including (1) scheduling, (2) powering-down nodes, (3) analyzing how HDFS data are processed, and (4) analyzing system components to address energy conservation.

Our proposed classification for energy-efficient MapReduce techniques (shown in Fig. 6) suggests dividing energy-efficient horizontal scaling techniques into two classes of homogeneous clusters versus heterogeneous clusters. Then each of these classes covers energy-efficient techniques for MapReduce programming model and scheduling, and for the DFS. Energy-efficient techniques proposed for homogenous clusters, mostly use scaling-down approach (i.e., transitioning servers to an inactive, low-power-consuming sleep/standby state) and scaling up when needed. Alternatively, in a heterogeneous MapReduce cluster can consist of high- and low-power machines. Also, modern machines are equipped with DVFS capabilities. These features provide opportunities to save energy by intelligently placing jobs and data on its corresponding energy-efficient machines.

4.1.1.1 Energy-Efficient MapReduce
In this section, we overview some of the energy-efficient MapReduce techniques for homogeneous and heterogeneous clusters. In homogeneous clusters, all machines are assumed to be the same; however, in the heterogeneous type, machines have different computing and consumption power.

All-in strategy (AIS) [70] is a framework for energy management in MapReduce clusters by powering down all nodes in the cluster during a low utilization period. It powers down all nodes during low utilization periods, batches the jobs and powers on whole nodes, performs all jobs, and again powers down all when entire jobs are completed. Due to its batch characteristics, AIS may not support an instant execution. However, it decreases the response time of workload by running the workload on all the available nodes in the cluster [44,70–72]. The AIS is difficult to be applied when MapReduce data analysis is interactive since the cluster is never to be completely inactive [44,71]. A method to cope with the problem of energy reduction in the presence of interactive jobs is to avoid replication [73]. This approach motivated by an empirical analysis of MapReduce interactive workload at Facebook. The analysis reveals that the interactive jobs operate on a small fraction of data; hence a small pool of machines (interactive zone) can be dedicated to these jobs. On the contrary, the less time-sensitive jobs can run on the rest of the cluster (batch zone) in a batch way. The interactive zone is always fully powered while the batch zone is powered down between batches. Consequently, energy saving comes from aggregating jobs in the batch zone.

Unlike the techniques that try to save energy by scaling down the nodes, some other approaches choose to run all nodes in medium power state to avoid the peak power states of servers [74]. This needs an energy-efficient job scheduling that can manage system workload so that few peaks observed in power consumption are removed. Further, a cluster can be dynamically reconfigured by scaling up and down the number of nodes based on the cluster utilization and current workload. The number of nodes is scaled up (down) in the cluster when the average cluster utilization rises above (falls below) a threshold specified by the cluster administrator. By doing this, the nodes in the cluster that are underutilized can be turned off to save power [75].

MapReduce efficiency may be varied depending on various factors in the system. Wirtz [68] consider variation of MapReduce efficiency with respect to two kinds of runtime configurations: resource allocation that changes the number of available concurrent workers, and DVFS that adjusts the processor frequency based on the workloads' computational needs. An optimization technique allocates the optimal number of compute nodes to applications and dynamically schedules processor frequency during its execution based on data movement characteristics.

Being aware of whether the system workloads are IO- or CPU-intensive a right strategy can be chosen for scheduling and mapping the tasks onto the server nodes. In a heterogeneous cluster, IO-bound workloads have better energy efficiency on low-power nodes while CPU-bound workloads achieve better energy efficiency on high-power nodes [76]. Further, the Map task is more CPU-intensive while a Reduce task is more IO-intensive. This observation leads to an energy-aware node selection for a given task. In

Ref. [77], a spatio-temporal trade-off is introduced that includes the efficient spatial placement of tasks on nodes and temporal placement of nodes with tasks having similar runtimes. The objective is to maximize utilization throughout its uptime for a datacenter with heterogeneous VMs. Based on the following two principles (1) efficient spatial fit and (2) balanced temporal fit, two VM placement algorithms are proposed. The former coplaces various MapReduce VMs based on their complementary CPU, memory, storage, and network requirements such that the available resources are fully utilized. The latter colocates MapReduce VMs with similar runtimes so that a server runs at a high utilization rate. Once all collocated VMs have finished, the cloud operator can hibernate the server to conserve energy. Having various VM types differing in the amount of CPU, memory, and storage capacity, each job is assigned a type of VM for its virtual cluster (from a set of predefined VM types).

It is not surprising to know that more parallelism does not always results in more energy efficiency or speedup, specifically when the job is I/O-intensive [78]. Network bandwidth is said to be a significant factor that affects the energy efficiency of the map-reduce jobs. Balancing CPU processing rate against I/O processing rate considering the workload characteristics and network bandwidth can improve the energy efficiency of map-reduce jobs.

DyScale [79] is a Hadoop scheduler that takes into account the heterogeneous features of the cores for achieving better performance objectives with a specified power budget. It enables "slow" slots (running on slow cores) and "fast" slots (running on fast cores) in Hadoop. Within the same power budget, DyScale operating on heterogeneous multicore processors provides significant performance improvement for small and interactive jobs comparing to using homogeneous processors with (many) slow cores. Moreover, DyScale maintains a good performance for large batch jobs compared to using a homogeneous fast core design (with fewer cores).

4.1.1.2 Energy-Efficient DFS

A solution for improving the energy efficiency of a DFS such as HDFS is to recast the data layout and task distribution of the file system to enable significant portions of a cluster to be powered down while still fully operational. Covering-set (CS) approach exploits the replication feature of DFSs. In CS strategy, some nodes in the system are specified as special nodes, called CS nodes. At least one copy of each unique data block is kept in these nodes. In this technique, during periods of low utilization, some or all of the non-CS nodes are powered down to save energy. The drawback of CS is that the workloads take longer to run when the cluster has been partially powered down, since fewer nodes are available to run the workload. It also requires code modifications in the underlying DFS software [71,72,80].

GreenHDFS [81] divides the cluster into Hot and Cold zones. Data are placed initially in the zones based on high-level data classification policies and later transitioned from one to another based on its access frequencies. The Hot zone consists of files that are being accessed currently and the newly created files. It includes high-performance, high-power, and high-cost CPUs. All nodes in Hot zone remain in high-power mode at all times. The majority of the servers in the cluster are assigned to Hot zone to ensure higher performance and higher data availability. The cold zone contains files with low to rare accesses. It utilizes machines with a higher number of disks to store a large amount of rarely accessed data. Aggressive power management policies are used in the cold zone to keep nodes in low, inactive power mode as long as possible. Nodes are powered-on only on demand [44,81].

4.1.2 Vertically Scaled Systems

Big data analytics including data mining is per essence parallel regardless of the programming model. Some different architectural designs and programming models have been introduced to speed up the execution of applications and improve the energy efficiency with massive data sets. Such platforms range from multicore clusters, hybrid clusters, clouds, mobile systems, GPUs, and FPGA systems. Also, a great attention is currently paid to substitute architectures for current server processors that will be discussed in the following sections.

4.1.2.1 Energy-Efficient Processing on GPUs

GPUs have been considered an excellent alternative to CPUs for high-performance and high-throughput computing applications. They can exploit data parallelism, increase computational efficiency, and save energy by orders of magnitude. An engine implemented in Refs. [33,82,83] is a query coprocessing engine on coupled CPU–GPU architecture (referred as APU) that can work as the base of in-memory database systems. This query coprocessing engine can improve the query processing performance by leveraging the hardware advances with optimized algorithm design. A fine-grained method distributes workload among available processors since the CPU, and the GPU share the same main memory space. Moreover, an in-cache paradigm is employed for query processing to take advantage of shared cache hierarchy to overcome memory stalls of query processing.

The comparison results of discrete CPU–GPUs and coupled CPU–GPUs show that the average power consumption of discrete architecture is between 36% and 44% higher than those of coupled architecture. The results indicate that discrete GPUs always deliver higher performance at the expense of more energy consumption. The coupled architectures win mainly due to the specifically designed architecture of APU so that various complicated interconnections are eliminated. Thus, it is more energy-efficient for database workloads.

4.1.2.2 Energy-Efficient FPGA-Based Processing

FPGAs can support very high rates of data throughput when high parallelism is exploited in circuits implemented in the reconfigurable fabric. FPGA reconfigurability offers a flexibility that makes them even superior to GPU for certain application domains. The key features of FPGA that can provide motivation for big data analytics are parallelism and efficient power consumption (performance/Watt). A vital feature of FPGA is its parallelism through a hierarchical style architecture that can be very much suitable for data processing applications. Many of the widely used and typical data operations can be implemented on FPGA through hardware programmability.

IBM's Netezza [84], which falls under data warehouse appliance category, is a big data infrastructure platform using FPGA. The analytics appliance includes custom-built FPGA accelerators. Netezza minimizes data movement by using innovative hardware acceleration. It employs FPGA to filter out extraneous data as early in the data stream as possible, and as fast as data can be streamed off the disk. Huge benefits by introducing FPGAs in big data analytics hardware have been proved. Specifically saying, the queries are compiled using FPGAs to minimize overhead. Each FPGA on server blades contains embedded engines that perform filtering and transformation functions on the data stream. These engines are dynamically reconfigurable that enables them to be modified or extended through software. They are customized for every snippet through instructions provided during query execution and act on the data stream at extremely high speeds.

LINQits [85] is a flexible and composable framework for accelerating data-intensive applications using specialized logic. LINQits accelerates a domain-specific query language called LINQ. It has been prototyped on a Xilinx Programmable SoC called the ZYNQ, which combines dual

ARM A9 processors and FPGA. LINQits improves energy efficiency by 8.9–30.6 times and performance by 10.7–38.1 times compared to optimized and multithreaded C programs running on conventional ARM A9 processors.

Academic researchers also showcased the vital development in building infrastructure for big data analytics. For example, BlueDBM or Blue Database Machine [99] is a storage system for big data analytics that can dramatically speed up the time it takes to access information. In this system, each inbuilt flash device is connected to FPGA chip to create an individual node. FPGAs are used not only to control the flash device but are also capable of performing processing operations on the data itself.

4.1.2.3 How to Choose Right Hardware?

A main challenge is to determine what type of hardware system is more suitable for which data type. It is quite essential for the processing infrastructure to be compute-intensive for running and processing such variety of data under time-bounded format. Most organizations with traditional data platforms such as enterprise data warehouses find that their existing infrastructure is either technically incapable or financially impractical for storing and analyzing big data. According to Intel [87], the selection of key specifications for a given application can be specific to each domain and operations involved in each system. Some of the most notable characteristics to help in decision making during the early stages of development are energy budget, bandwidth, data width, read/write speed, data process mechanism (e.g., batch and stream processing), network switching topology, traffic density, congestion control, and security standards. All these characteristics define four distinctive categories of system specifications for hardware infrastructure including compute, memory, storage, and network.

GPU and FPGA are the possible accelerators that can achieve higher performance and energy efficiency than CPUs on certain jobs. GPU is power-efficient but only for SIMD (single instruction multiple data) streams, and the FPGA is hard to program. However, the data-flow style architecture in FPGA may dominate CPU and GPU in providing high-performance memory-intensive operations at low-power consumption (performance/Watt) for a category of operations. The major overhead of GPUs and CPUs comes from executing instructions that rely on memory accesses. FPGA took the advantage of data-flow streaming, thus saving many of the memory accesses. The main drawback with FPGA is the programming complexity. Although, big data processing is performance-intensive, some applications specifically require reduced energy cost. According to Refs. [23, 33], FPGA could be a viable solution on an energy cost basis for very high-performance, large-scale applications compared to GPU and CPU.

Although FPGA is a winner with respect to the power efficiency, still GPU is outperforming its counterparts in terms of performance for some applications such as multimedia and communication algorithms from the HPC domain [88] that often make extensive use of floating-point arithmetic operations. Due to the fact that complexity and expense of the floating-point hardware on a reconfigurable fabric such as FPGA are high, these algorithms are converted to fixed-point operations thus making FPGA less efficient than GPU for achieving higher speeds.

It is worth mentioning that a combination of FPGA, GPU, and CPU hardware infrastructure is giving good results for the applications such as medical imaging [89]. Even though it tends to be very expensive to develop such true heterogeneous infrastructure, the choice is purely based on the user requirement. Another significant requirement of heterogeneity is in signal processing domain that is in need of signal filtering and transformation, encoding/decoding, floating-point operations, etc.

4.2 New Server Architectures

Recent research shows that the architectures of current servers do not comply well the computational requirements of big data processing applications. Therefore, it is required to look for a new architecture for servers as a replacement for currently used machines for both performance and energy enhancement. Using low-power processors (microservers), more system-level integration and a new architecture for server processors are some of the solutions that have been discussed recently as performance/energy-efficient replacement for current machines.

4.2.1 Microservers for Big Data Analytics

Processing clusters based on low-power systems such as ARM processors are feasible, and this is an appropriate way to decrease power usage of several server applications. Prior research shows that the processors based on simple in-order cores are well suited for certain scale-out workloads [63]. A comparison between ×86 and ARM architectures for server applications [31] concludes that ARM-based processors are three to four times more energy-efficient than the ×86-based processors while comparing requests per second per Watt relation [31,60].

Several other studies have shown the efficiency of ARM microservers. One of these studies investigates the energy performance of server workloads on ARM Cortex-A9 multicore system through a trace-driven analytical model [62]. The model involves the degrees of CPU core, memory, and I/O resource overlap and estimates the optimal number of cores and clock frequency in favor of higher energy efficiency without compromising execution time. Using the collected metrics and a static power profile of the system, the execution time and energy usage of an application is predicted for a various number of cores and clock frequencies. Then, the configuration that maximizes performance without wasting energy is selected. It is predicted that the selected configurations increase energy efficiency by 10% without turning off cores and just through frequency scaling, and up to one-third with shutting down unutilized cores. For memory-bounded programs, the limited memory bandwidth might increase both execution time and energy usage to the point where energy cost might be higher than on a typical ×64 multicore system. The conclusion is that increasing memory and I/O bandwidth can improve the execution time and the energy usage of server workloads on ARM Cortex-A9 systems. Moreover, a 3000-node cluster simulation driven by a real-world trace from Facebook shows that on average a cluster comprising ARM-based microservers which support the Hadoop platform reaches the same performance of standard servers while saving energy up to 31% at only 60% of the acquisition cost.

Recently, ARM big.LITTLE boards (as small nodes) have been introduced as a platform for big data processing [61]. In comparison with Intel Xeon server systems (as traditional big nodes), the I/O-intensive MapReduce workloads are more energy-efficient to run on Xeon nodes. In contrast, database query processing is always more energy-efficient on ARM servers, at the cost of slightly lower throughput. With minor software modifications, CPU-intensive MapReduce workloads are almost four times cheaper to execute on ARM servers. Unfortunately, small memory size, low memory, and I/O bandwidths, and software immaturity ruins the lower-power advantages obtained by ARM servers. Efficient employment of ARM servers while coping with existing architectural constraints may be considered for further research and development.

4.2.2 Novel Server Processors

It has been shown that scale-out workloads have many characteristics that need to be known as a distinct workload class from desktop, parallel, and traditional server workloads [56]. Due to the large mismatch between the demands of the scale-out workloads and today's processor micro-architecture, scale-out processors have been recently introduced that can result in more area- and energy-efficient servers in future [56,64,65]. The building block of a scale-out processor is the pod. A pod is a complete server that runs its copy of the OS. A pod acts as the tiling unit in a scale-out processor, and multiple pods can be placed on a die. A scale-out chip is a simple composition of one or more pods and a set of memory and I/O interfaces. Each pod couples a small last-level cache to a number of cores using a low-latency interconnect. Having a higher per-core performance and lower energy per operation leads to better energy efficiency in scale-out processors. Due to smaller caches and smaller communication distances, scale-out processors dissipate less energy in the memory hierarchy [65].

FAWN architecture [45] is another solution for building cluster systems for energy-efficient serving massive-scale I/O and data-intensive workloads. FAWN couples low-power and efficient embedded processors with flash storage to provide fast and energy-efficient processing of random read-intensive workloads.

4.2.3 System-Level Integration (Server-on-Chip)
System-level integration is an alternative approach that has been proposed for improving the efficiency of the warehouse-scale datacenter server market. System-level integration discusses placing CPUs and components on the same die for servers, as done for embedded systems. Integration reduces the (1) latency: by placing cores and components closer to one another, (2) cost: by reducing parts in the bill of material, and (3) power: by decreasing the number of chip-to-chip pin-crossings. Initial results show a reduction of more than 23% of capital cost and 35% of power costs at 16 nm [11].

5. TRENDS FOR THE BIG DATA ANALYTICS IN CLOUD-BASED DATACENTERS

Despite great changes in the way data are accessed and managed, the datacenter has kept its server-centric model which is not adequate to keep up with new services, new energy requirements, and new business models [4]. To accommodate big data analytics, the next-generation cloud-based datacenter should comply with a new set of requirements. In the following, the challenges and opportunities in big data analytics are discussed for future research and development.

5.1 Energy Efficiency

Measurements on current datacenters indicate significant wasted energy due to underutilization with average utilization values of about 30% [90]. Today's application environments are more distributed, often with multiple tiers, and oriented toward service delivery, eventually resulted in:

- Greater traffic volume on the Ethernet network,
- more storage traffic as applications use DFSs and increase the amount of synchronization and replication data across the network,
- greater traffic flow between peer nodes such as server-to-server or VM-to-VM [20].

The massive number of processing units puts interconnection network systems under pressure to increase performance and accommodates better the communication among them [91]. The higher level of network complexity and higher link bandwidth increase the energy consumed, and the heat emitted. The energy cost and the heat dissipation problems in interconnection networks necessitate future network systems to be built with much more efficient power than today. Furthermore, computing with renewable energy is crucial in the future datacenters. There is a strong need for innovative technologies for improved energy capturing, and management of energy usage with energy storage and availability [4].

Table 2 shows the projection of performance, bandwidth, and power consumption for future datacenters. As one can see, the peak performance and bandwidth requirements are highly demanding while the affordable power consumption is in the much lower increasing rate. This reveals the great challenges in power consumption of future datacenters and the opportunities for new research in this area. Recently, energy proportionality received some attention as a viable solution for power consumption in datacenters [12], though need deeper investigation to be utilized in next-generation datacenters.

Table 2 Performance, Bandwidth Requirements, and Power Consumption Bound for Future Systems [92]

Year	Peak Performance (10× Every 4 years; PFLOPS)	Bandwidth Requirements (20× Every 4 years; PB/s)	Power Consumption Bound (2× Every 4 years; MW)
2012	10	1	5
2016	100	20	10
2020	1000	400	20

5.2 Interconnection Networks

To support the big data analytics in datacenters, the high-throughput/bandwidth interconnection networks with low latency and low-power consumption are essential. This includes the data and storage area networks (e.g., SAN) as by increasing the size of the data the demand rise in storage systems as well. Depending on the kind of data management software in use and the kinds of data analyzed, big data can influence the size and frequency of data movements among servers and between servers and storage [86]. Moving large objects (e.g., videos or high-resolution images) from long-term storage to analytical nodes will place a bigger burden on storage systems and data networks than moving myriad small text files [3,54]. Sizing network links, prioritization, traffic shaping, and data compression can be remedies [4,54]. Due to increasing rate of data-intensive applications requiring ubiquitous connectivity, 400G Ethernet will be required in datacenters/routers. Cost considerations are a main reason that copper wires are still favorable even for high-bandwidth networks [93]. Moreover, using optical technologies is a viable solution that needs further research and investigation for future DCNs.

5.3 Datacenters Architecture

Distributing data-intensive applications across multiple datacenters seems to be a good solution for efficient big data management in the clouds. However, it should be supported by a reliable framework for managing large and distributed data sets and mobile databases stored on mobile devices over multiple datacenters while optimizing the network (e.g., latency and throughput) and cloud service QoS (e.g., cost, response time, etc.) [4,20]. There is a need for following technologies to realize distributed applications over multiple datacenters:

- Collaborative distributed datacenters that can support both location- and workload-dependent services and redundancy for failure recovery. The system architecture should be based on multiple datacenters, including scheduling, power management, and data/workload placement.
- Innovative networking fabrics and distributed systems to extend services across multiple heterogeneous datacenters.
- Optimal resource provisioning across multiple datacenters. Large-scale distributed data-intensive applications, e.g., environment monitoring applications, need to process and manage massive data sets across geographically distributed datacenters. These kinds of applications that

combine multiple, independent, and geographically distributed software and hardware resources require provisioning across multiple datacenter resources.

- Software frameworks and services that allow portability of distributed applications across multiple datacenters. Despite the existing technological advances of the data processing paradigms (e.g., MapReduce and cloud computing platforms), large-scale, reliable system-level software for big data applications is yet to become commonplace.
- Appropriate programming abstractions, which can extend the capability of existing data processing paradigms to multiple datacenters.

5.3.1 Heterogeneity and Virtualization

The data and processing platform heterogeneity is inevitable in the future of the datacenters. The concept of datacenter remains the same as before, but the physical implementation will be different from the prior generations. The next-generation datacenter will be deployed on a heterogeneous infrastructure and architectures that integrate both traditional structured data and big data into one scalable environment [21]. New server and processor architectures appropriately matching with big data analytics requirements pose an indispensable trend for the microarchitects. Virtualization infrastructure is the key concept in datacenters to support hardware and software heterogeneity and simplify the resource provisioning [3]. Virtualization has a wide concept and has been studied for several years for datacenters and cloud computing systems. The high-level language virtualization is the most relevant topic for big data analytics, which allow languages to be executed on multiple computing architectures. This needs to be done through compiler and runtime environment (e.g., JVM). The main challenge here is to consider the programmer's productivity in future big data analytics software since it has a great impact on system cost.

5.4 Resource Provisioning for Big Data Applications

Processing of uncertain and heterogeneous data volumes requires optimal provisioning of cloud-based datacenters. The process of provisioning hardware and software resources to big data applications requires the resource provider to compute the best software and hardware configuration to ensure that QoS targets of applications are achieved, while maximizing energy efficiency and utilization. Schad *et al.* [94] show that the QoS uncertainty of applications is the main technical obstacle to the successful adoption of cloud datacenters.

The uncertainties of resource provisioning have two aspects. First, from the perspective of big data applications, it is difficult to estimate its workload behavior in terms of data volume to be analyzed, data arrival rate, data types, data processing time distributions, I/O system behavior, number of users, and type of network connecting to the application. Second, from a resource perspective, without knowing the behaviors of big data applications, it is difficult to make decisions about the size of resources to be provisioned at a certain time. Furthermore, the availability, load, and throughput of datacenter resources can vary in unpredictable ways, due to failure, malicious attacks, or congestion of network links. In other words, we need reasonable application workload and resource performance prediction models when making provisioning decisions [20].

One possible way to deal with uncertainties is to rely on monitoring the state of both hardware and software resources and taking predefined actions when some events occur. In Ref. [95], the OS-level metrics such as available CPU percentage, available non-paged memory, and TCP/IP performance are gathered for such purpose. A network QoS-aware provisioning of MapReduce framework for private cluster computing environments has been introduced in Ref. [24]. However, this problem is more challenging and poses many open problems on the QoS-based predictive resource provisioning for multiple cloud datacenters [19]. Therefore, there are several challenges in the resource provisioning that should be researched to provide QoS requirements of big data applications.

6. SUMMARY

Highly virtualized cloud-based datacenters currently provide the main platform for big data analytics. As the scale of datacenters is increasingly expanding to accommodate big data needs, minimizing energy consumption and operational cost is a vital concern. The datacenters infrastructure including interconnection network, storage, and servers should be able to handle big data applications in an energy-efficient way. The datacenter may be vertically scaled and equipped with more and faster hardware, processors, and memory to be able to handle large future workloads. Furthermore, horizontal scaling distributes the work across many servers which is an economical way to increases the performance compared with vertical scaling. Improving the utilization rate of servers and network resources, adjusting routing to the traffic flow, using optical interconnection networks, consolidating applications through virtualization in the cloud, optimizing network topologies established among VMs, efficient distribution of work-load in the federated multicloud datacenters, and optimizing the power consumption of servers, processors, power supplies, and memory system are some of techniques for energy-efficient data analytics in datacenters. There are still many challenges and open problems as well as the need for improved or new technologies in the architecture and interconnection networks, resource provisioning, etc.

REFERENCES

[1] Hewlett-Packard Technical White Paper, Powering the 21st-century datacenter with HP next generation FlexFabric, Hewlett-Packard, Technical white paper, 4AA4-6632ENW, p. 1–12 May 2013.

[2] P. Castoldi, N. Androiolli, I. Cerutti, O. Liboiron-Ladouceur, P.G. Raponi, Energy efficiency and scalability of multi-plane optical interconnection networks for computing platforms and datacenters. in: Optical Fiber Communication Conference and Exposition (OFC/NFOEC), 2012. http://dx.doi.org/10.1364/OFC.2012.OW3J.4.

[3] M. Al-Fares, A. Loukissas, A. Vahdat, A scalable, commodity datacenter network architecture, in: Proceedings of the ACM SIGCOMM Conference on Data Communication, 2008, pp. 63–74.

[4] G. Cook, How clean is your cloud?: Report, Morgan and Claypool Publishers, 2012. July 2013, 154 pages, Available at: http://www.morganclaypool.com/doi/abs/10.2200/S00516ED2V01Y201306CAC024.

[5] C. Wang, Survey of Recent Research Issues in Datacenter Networking, 2013)(online) http://www.cse.wustl.edu/~jain/cse570-13/ftp/dcn.

[6] L.A. Barroso, J. Clidaras, U. Holzle, The Datacenter as a Computer: An Introduction to the Design of Warehouse-Scale Machines, second ed., Morgan & Claypool Publishers, 2013. ISBN: 9781627050098.

[7] C. Kachris, K. Bergman, I. Tomkos, Optical Interconnects for Future Datacenter Networks, Springer-Verlag, New York, 2012.

[8] N. Farrington, G. Porter, S. Radhakrishnan, H. Bazzaz, V. Subramanya, Y. Fainman, G. Papen, A. Vahdat, Helios: a hybrid electrical/optical switch architecture for modular data centers, in: Proceedings of ACM SIGCOMM Computer Communication Review, vol. 40, 2010, pp. 339–350.

[9] A. Lakshman, P. Malik, Cassandra: structured storage system on a p2p network, in: Proceedings of the 28th ACM Symposium on Principles of Distributed Computing, 2009.

[10] U. Hoelzle, L.A. Barroso, The Datacenter as a Computer: An Introduction to the Design of Warehouse-Scale Machine, first ed., Morgan & Claypool Publishers, 2009. ISBN: 9781598295566.

[11] S. Li, K. Lim, P. Faraboschi, J. Chang, P. Ranganathan, N.P. Jouppi, System-level integrated server architectures for scale-out datacenters, in: Proceedings of the 44th Annual IEEE/ACM International Symposium on Microarchitecture, 2011.

[12] GreenDataProject, Where Does Power Go? 2008 (online). http://www.greendataproject.org.

[13] C. Guo, H. Wu, K. Tan, L. Shi, Y. Zhang, S. Lu, DCell: a scalable and fault-tolerant network structure for datacenters, in: Proceedings of ACM SIGCOMM Computer Communication Review, vol. 38, issue 4, 2008, pp. 75–86.

[14] C. Guo, G. Lu, D. Li, H. Wu, X. Zhang, Y. Shi, C. Tian, Y. Zhang, S. Lu, BCube: a high performance, server-centric network architecture for modular datacenters,

in: Proceedings of ACM SIGCOMM Computer Communication Review, vol. 39, issue 4, 2009, pp. 63–74.

[15] G. Wang, D. Andersen, M. Kaminsky, K. Papagiannaki, T. Ng, M. Kozuch, M. Ryan, c-Through: part-time optics in data centers, in: Proceedings of ACM SIGCOMM Computer Communication Review, vol. 40, 2010, pp. 327–338.

[16] K. Chen, A. Singla, A. Singh, K. Ramachandran, L. Xu, Y. Zhang, X. Wen, Y. Chen, OSA: an optical switching architecture for data center networks with unprecedented flexibility, in: Proceedings of the 9th USENIX Conference on Networked Systems Design and Implementation, 2012.

[17] J. Dean, S. Ghemawat, MapReduce: simplified data processing on large clusters, Commun. ACM 51 (1) (2008) 107–113.

[18] Hewlett-Packard Development Company, Effects of Virtualization and Cloud Computing on Datacenter Networks. Technology Brief, 2011.

[19] B. Di-Martino, R. Aversa, G. Cretella, A. Esposito, Big data (lost) in the cloud, Int. J. Big Data Intelligence 1 (1/2) (2014) 3–17.

[20] G. DeCandia, et al., Dynamo: Amazon's highly available key-value store, ACM SIGOPS Oper. Sys. Rev. 41 (6) (2007) 205–220.

[21] IBM Infosphere BigInsights. http://www.ibm.com/software/data/infosphere/biginsights/, 2013 (online).

[22] M. Saecker, V. Markl, Big data analytics on modern hardware architectures: a technology survey, business intelligence, Lect. Notes Bus. Inf. Process. 138 (2013) 125–149.

[23] K.C. Nunna, F. Mehdipour, A. Trouve, K. Murakami, A survey on big data processing infrastructure: evolving role of FPGA, Int. J. Big Data Intelligence 2 (3) (2015) 145–156. http://www.inderscience.com/info/ingeneral/forthcoming.php?jcode=IJBDI.

[24] N. Rizvandi, J. Taheri, R. Moraveji, A.Y. Zomaya, Network load analysis and provisioning of MapReduce applications, in: Proceedings of International Conference on Parallel and Distributed Computing, Applications, and Technologies (PDCAT), 2013.

[25] K. Krishnan, Data Warehousing in the Age of Big Data, Morgan Kaufmann, Waltham, MA, 2013. ISBN: 978-0-12-405891-0.

[26] C. Patel, C. Bash, R. Sharma, M. Beitelmam, R. Friedrich, Smart cooling of datacenters, in: Proceedings of InterPack, July, 2003.

[27] J. Anderson, J. Lehnardt, N. Slater, CouchDB: The Definitive Guide, O'Reilly Media, Inc., Sebastopol, CA, 2010.

[28] Opani. http://www.opani.com, 2013 (online).

[29] Infochimps. http://www.infochimps.com, 2013 (online).

[30] S. Nedevschi, J. Chandrashekar, B. Nordman, Skilled in the art of being idle: reducing energy waste in networked systems, in: Proceedings of the 6th USENIX Symposium on Networked Systems Design and Implementation, April, 2009, pp. 381–394.

[31] R.V. Aroca, L.M. Garcia Gonçalves, Towards green data centers: a comparison of x86 and ARM architectures power efficiency, J. Parallel Distrib. Comput. 72 (2012) 1770–1780.

[32] J. Armstrong, OFDM for optical communications, J. Lightwave Technol. 27 (3) (2009) 189–204.

[33] T. Hamada, et al., A comparative study on ASIC, FPGAs, GPUs and general purpose processors in the O(N 2) gravitational N-body simulation, in: Proceedings of NASA/ESA Conference on Adaptive Hardware and Systems, 2009.

[34] M. McKusick, S. Quinlan, GFS: evolution on fast-forward, Queue 7 (7) (2009) 10.

[35] L. Huff, The choice for datacenter cabling, in: Berk-Tek Technology Summit, 2008.

[36] D. Beaver, et al., Finding a needle in haystack: Facebook's photo storage, in: OSDI, vol. 10, 2010.

[37] R. Steinmetz, K. Wehrle, Peer-to-Peer Systems and Applications, Springer, Berlin and Heidelberg, 2005.

[38] F. Chang, et al., Bigtable: a distributed storage system for structured data, ACM Trans. Comput. Sys. 26 (2) (2008) 4.

[39] D.S. Milojicic, V. Kalogeraki, R. Lukose, K. Nagaraja, J. Pruyne, B. Richard, S. Rollins, Z. Xu, Peer-to-peer computing: Technical report HPL-2002-57, HP Labs, 2002.

[40] D. Singh, C.K. Reddy, A survey on platforms for big data analytics, J. Big Data 2 (2014) 8.

[41] C. Olston, B. Reed, U. Srivastava, R. Kumar, A. Tomkins, Pig latin: a not-so-foreign language for data processing, in: Proceedings of the ACM SIGMOD International Conference on Management of Data, 2008, pp. 1099–1110.

[42] A. Thusoo, et al., Hive: a warehousing solution over a map-reduce framework, in: Proceedings of the VLDB Endowment, vol. 2, issue 2, 2009, pp. 1626–1629.

[43] K. Shvachko, et al., The Hadoop distributed file system, in: Proceedings of the IEEE 26th Symposium on Mass Storage Systems and Technologies (MSST), 2010.

[44] N. Tiwari, Scheduling and energy efficiency improvement techniques for Hadoop map-reduce: state of art and directions for future research, Ph. D. Seminar report, Department of Computer Science and Engineering, Indian Institute of Technology, Bombay, India, 2012.

[45] D.G. Andersen, et al., FAWN: a fast array of wimpy nodes, in: SOSP'09, 2009, pp. 1–14.

[46] R. Barga, J. Ekanayake, W. Lu, Project Daytona: data analytics as a cloud service, in: IEEE 28th International Conference on Data Engineering (ICDE), IEEE, 2012.

[47] H. Herodotou, et al., Starfish: a self-tuning system for big data analytics, in: 5th Biennial Conference on Innovative Data Systems Research (CIDR'11), 2011, pp. 261–272.

[48] H. Herodotou, S. Babu, Profiling, what-if analysis, and cost-based optimization of MapReduce programs, in: Proceedings of the VLDB Endowment, vol. 4, issue 11, 2011, pp. 1111–1122.

[49] M. Zaharia, M. Chowdhury, M.J. Franklin, S. Shenker, I. Stoica, Spark—cluster computing with working sets, in: Proceedings of the 2nd USENIX Conference on Hot Topics in Cloud Computing, 2010.

[50] I. Stoica, Berkeley Data Analytics Stack (BDAS) Overview, O'Reilly Strata Conference, Santa Clara, CA, Feb. 26–28, 2013.

[51] M. Isard, et al., Dryad: distributed data-parallel programs from sequential building blocks, ACM SIGOPS Oper. Sys. Rev. 41 (3) (2007).

[52] G. Malewicz, et al., Pregel: a system for large-scale graph processing, in: Proceedings of the ACM SIGMOD International Conference on Management of data, 2010.

[53] R. Buyya, High Performance Cluster Computing: Architectures and Systems, vol. 1, Prentice Hall, Upper Saddle River, NJ, 1999.

[54] K.S. Beyer, V. Ercegovac, R. Gemulla, A. Balmin, M. Eltabakh, C.C. Kanne, F. Ozcan, E.J. Shekita, Jaql: a scripting language for large scale semistructured data analysis, in: The Proceedings of the VLDB Endowment (PVLDB), 2011, pp. 1272–1283.

[55] D. Battre, S. Ewen, F. Hueske, O. Kao, V. Markl, D. Warneke, Nephele/PACTs: a programming model and execution framework for Web-scale analytical processing, in: Proceedings of the 1st ACM Symposium on Cloud Computing, SoCC, 2010, pp. 119–130.

[56] M. Ferdman, et al., Clearing the clouds: a study of emerging scale-out workloads on modern hardware, in: 17th International Conference on Architectural Support for Programming Languages and Operating Systems (ASPLOS), March, 2012.

[57] D.J. DeWitt, R.H. Gerber, G. Graefe, M.L. Heytens, K.B. Kumar, M. Muralikrishna, GAMMA—a high performance dataflow database machine, in: Proceedings of the 12th International Conference on Very Large Data Bases (VLDB), Morgan Kaufmann Publishers Inc., San Francisco, CA, 1986, pp. 228–237.

[58] Teradata Corporation, Teradata, 2011 (online). http://www.teradata.com/.

[59] A. Anwar, K.R. Krish, A.R. Butt, On the use of microservers in supporting Hadoop applications, in: IEEE International Conference on Cluster Computing (CLUSTER), 2014.

[60] J. Shuja, K. Bilal, S.A. Madani, M. Othman, R. Ranjan, P. Balaji, S.U. Khan, Survey of techniques and architectures for designing energy-efficient data centers, IEEE Sys. J. 99 (2014) 1–13.

[61] D. Loghin, B.M. Tudor, H. Zhang, B.C. Ooi, Y.M. Teo, A performance study of big data on small nodes, in: Proceedings of the VLDB Endowment, vol. 8, issue 7, 2015.

[62] B.M. Tudor, Y.M. Teo, On understanding the energy consumption of ARM-based multicore servers, in: Proceedings of the 34th ACM SIGMETRICS, Pittsburgh, USA, 2013.

[63] K. Lim, P. Ranganathan, J. Chang, C. Patel, T. Mudge, S. Reinhardt, Understanding and designing new server architectures for emerging warehouse-computing environments, in: Proceedings of the International Symposium on Computer Architecture, June, 2008.

[64] N. Hardavellas, I. Pandis, R. Johnson, N. Mancheril, A. Ailamaki, B. Falsafi, Database servers on chip multiprocessors: limitations and opportunities, in: The Conference on Innovative Data Systems Research, January, 2007.

[65] P. Lotfi-Kamran, et al., Scale-out processors, in: 39th Annual International Symposium on Computer Architecture International Symposium on Computer Architecture (ISCA), 2012.

[66] G. Magklis, M. Scott, G. Semeraro, et al., Profile-based dynamic voltage and frequency scaling for a multiple clock domain microprocessor, June, in: ISCA'03, 2003.

[67] CALIENT Optical Switching in the New Generation Datacenter. http://www.telecomreviewna.com/index.php?option=com_content&view=article&id=151:calient-optical-switching-in-the-new-generation-datacenter&catid=40:march-april-2012&Itemid=80 (online).

[68] T.S. Wirtz, Energy Efficient Data-Intensive Computing with Mapreduce, Master's Theses (2009). Paper 211. http://epublications.marquette.edu/theses_open/211, 2013.

[69] N. Vasic, M. Barisits, V. Salzgeber, D. Kostic, Making cluster applications energy-aware, in: ACM Proceedings of the 1st Workshop on Automated Control for Datacenters and Clouds (ACDC), 2009, pp. 37–42.

[70] W. Lang, J.M. Patel, Energy management for MapReduce clusters, in: Proceedings of the VLDB Endowment, vol. 3, issue 1–2, 2010, pp. 129–139.

[71] A. Elsayed, O. Ismail, M.E. El-Sharkawi, MapReduce: state-of-the-art and research directions, Int. J. Comput. Electr. Eng. 6 (1) (2014) 34–39.

[72] K.H. Lee, Y.J. Lee, H. Choi, Y.D. Chung, B. Moon, Parallel data processing with MapReduce: a survey, Newsl. ACM SIGMOD Rec. Arch. 40 (4) (2011) 11–20.

[73] Y. Chen, S. Alspaugh, D. Borthakur, R. Katz, Energy efficiency for large-scale mapreduce workloads with significant interactive analysis, in: Proceedings of the 7th ACM European Conference on Computer Systems, 2012, pp. 43–56.

[74] N. Zhu, L. Rao, X. Liu, J. Liu, H. Guan, Taming power peaks in MapReduce clusters, in: Proceedings of the ACM SIGCOMM Conference, 2011, pp. 416–417.

[75] N. Maheshwari, R. Nanduri, V. Varma, Dynamic energy-efficient data placement and cluster reconfiguration algorithm for MapReduce framework, Futur. Gener. Comput. Sys. 28 (1) (2012) 119–127.

[76] N. Yigitbasi, K. Datta, N. Jain, T. Willke, Energy-efficient scheduling of MapReduce workloads on heterogeneous clusters, in: Proceedings of the 2nd International Workshop on Green Computing Middleware, 2011, pp. 1–6.

[77] M. Cardosa, A. Singh, H. Pucha, A. Chandra, Exploiting spatio-temporal tradeoffs for energy-aware MapReduce in the Cloud, IEEE Trans. Comput. (2012) 1737–1751.

[78] N. Tiwari, S. Sarkar, U. Bellur, M. Indrawan, An empirical study of Hadoop's energy efficiency on a HPC cluster, in: 14th International Conference on Computational Science (ICCS), 2014.

[79] F. Yan, L. Cherkasova, Z. Zhang, E. Smirni, DyScale: a MapReduce Job scheduler for heterogeneous multicore processors, IEEE Trans. Cloud Comput. 99 (2015).

[80] J. Leverich, C. Kozyrakis, On the energy (in) efficiency of Hadoop clusters, ACM SIGOPS Oper. Sys. Rev. 44 (1) (2010) 61–65.

[81] R.T. Kaushik, M. Bhandarkar, K. Nahrstedt, Evaluation and analysis of GreenHDFS: a self-adaptive, energy-conserving variant of the Hadoop distributed file system, in: Proceedings of 2nd IEEE International Conference on Cloud Computing Technology and Science, 2010, pp. 274–287.

[82] S. Zhang, J. He, B. He, M. Lu, OmniDB: towards portable and efficient query processing on parallel CPU/GPU architectures, in: Proceedings of the VLDB Endowment, vol. 6, issue 12, 2013, pp. 1374–1377.

[83] J. He, S. Zhang, B. He, In-cache query co-processing on coupled CPU-GPU architectures, in: Proceedings of the VLDB Endowment, vol. 8, issue 4, 2014, pp. 329–340.

[84] P. Francisco, The Netezza data appliance architecture: a platform for high performance data warehousing and analytics, in: IBM Redbooks, 2011.

[85] E.S. Chung, J.D. Davis, J. Lee, LINQits: big data on little clients, in: 40th Annual International Symposium on Computer Architecture, 2013.

[86] B. Javadi, B. Zhang, M. Taufer, Bandwidth modeling in large distributed systems for big data applications, in: The 15th International Conference on Parallel and Distributed Computing, Applications and Technologies, Hong Kong, 2014.

[87] Intel Big Data Analytics White Paper, extract, transform and load big data with Apache Hadoop, 2013.

[88] C. Cullinan, et al., Computing performance benchmarks among CPU, GPU, and FPGA, MathWorks (2012), 124 pages.

[89] P. Meng, et al., FPGA-GPU-CPU heterogenous architecture for real-time cardiac physiological optical mapping, in: Proceedings of the International Conference on Field-Programmable Technology, 2012.

[90] H. Wang, K. Bergman, Optically interconnected datacenter architecture for bandwidth intensive energy-efficient networking, in: Proceedings of 14th International Conference on Transparent Optical Networks (ICTON), 2012.

[91] H. Nguyen, D. Franco, E. Luque, Performance-aware energy saving mechanism in interconnection networks for parallel systems, in: ICCS 2014. 14th International Conference on Computational Science, vol. 29, 2014, pp. 134–144.

[92] A. Benner, Optical interconnect opportunities in supercomputers and high end computing, in: Optical Fiber Communication Conference, Optical Society of America, 2012.

[93] R. Navid, High-speed and power-efficient serial links for big data, in: International Workshop on IoT Enabling Chips, June, Kyoto Institute of Technology, Japan, 2015.

[94] J. Schad, J. Dittrich, J.A. Quiane-Ruiz, Runtime measurements in the cloud: observing, analyzing, and reducing variance, in: Proceedings of the VLDB Endowment, vol. 3, issue 1–2, 2010, pp. 460–471.

[95] R. Wolski, Experiences with predicting resource performance on-line in computational grid settings, ACM SIGMETRICS Perform. Eval. Rev. 30 (4) (2009) 41–49.

[96] Amazon Elastic MapReduce (Amazon EMR). http://aws.amazon.com/elasticmapreduce/.

[97] S. Fushimi, M. Kitsuregawa, H. Tanaka, An overview of the system software of a parallel relational database machine GRACE, in: Proceedings of the 12th International Conference on Very Large Data Bases (VLDB), Morgan Kaufmann Publishers Inc., San Francisco, CA, 1986, pp. 209–219.

[98] N. Ghani, S. Dixit, T.S. Wang, On IP-over-WDM integration, IEEE Commun. Mag. 38 (3) (2000) 72–84.

[99] S.W. Jun, et al., Scalable multi-access flash store for big data Analytics, in: Proceedings of the International Conference on Field-Programmable Gate Arrays, 2014.

[100] Y. Liu, J.K. Muppala, M. Veeraraghavan, D. Lin, M. Hamdi, Datacenter Networks—Topologies, Architectures and Fault-Tolerance Characteristics, Springer, 2013 (ISSN 2191–5768).

[101] B. Peer, P. Rajbhoj, N. Chathanur, Complex events processing: unburdening big data complexities, Infosys Labs Briefings 11 (1) (2013).

[102] PiCloud. http://www.picloud.com/, 2013 (online).

ABOUT THE AUTHORS

Farhad Mehdipour is an Associate Professor at E-JUST Center of Kyushu University, Fukuoka, Japan. He was a Research Associate at the School of Information Science and Electrical Engineering, Kyushu University during December 2006 to July 2010. He received his B.Sc. degree from Sharif University of Technology in 1996, and the M.Sc. and Ph.D. degrees in Computer Systems Architectures from the Amirkabir University of Technology in 1999 and 2006, respectively. He served as the organizing and program committee member or invited speaker in many international conferences and workshops. His research interests are high-performance and low-power micro-architectures, data analytics, and enabling technologies for cyber-physical systems including wireless sensor networks and IoTs. He is a senior member of IEEE.

Hamid Noori is an Assistant Professor at Ferdowsi University of Mashhad (FUM), Faculty of Engineering, Electrical and Computer Engineering Division. Before joining FUM at January 2011, he was an Assistant Professor at University of Tehran, College of Engineering, School of Electrical and Computer Engineering, during January 2009 to January 2011. He received his B.Sc., M.Sc., and PhD degrees all in Computer Engineering from the Sharif University of Technology (Iran) in 1996, Amirkabir University of Technology (Iran) in 2000, and Kyushu University (Japan) in 2007, respectively. His research interests include parallel processing, multi-core processors, multi-processor, and embedded systems.

Bahman Javadi is a Senior Lecturer in Networking and Cloud Computing at the Western Sydney University, Australia. Prior to this appointment, he was a Research Fellow at the University of Melbourne, Australia. From 2008 to 2010, he was a Postdoctoral Fellow at the INRIA Rhone-Alpes, France. He received his M.S. and Ph.D. degrees in Computer Engineering from the Amirkabir University of Technology in 2001 and 2007, respectively. He has been a Research Scholar at the School of Engineering and Information Technology, Deakin University, Australia during his Ph.D. course. He is co-founder of the Failure Trace Archive, which serves as a public repository of failure traces and algorithms for distributed systems. He has received numerous Best Paper Awards at IEEE/ACM conferences for his research papers. He served as a program committee of many international conferences and workshops. His research interests include Cloud and Grid computing, performance evaluation of large scale distributed computing systems, and reliability and fault tolerance. He is a member of ACM and senior member of IEEE.

CHAPTER THREE

Energy-Efficient and SLA-Based Resource Management in Cloud Data Centers

Altino M. Sampaio*, Jorge G. Barbosa†

*CIICESI, Instituto Politécnico do Porto, Escola Superior de Tecnologia e Gestão de Felgueiras, Felgueiras, Portugal
†Faculdade de Engenharia, Departamento de Engenharia Informática, LIACC, Universidade do Porto, Porto, Portugal

Contents

Advances in Computers, Volume 100
ISSN 0065-2458
http://dx.doi.org/10.1016/bs.adcom.2015.11.002

103

Abstract

Nowadays, cloud data centers play an important role in modern Information Technology (IT) infrastructures, being progressively adopted in different scenarios. The proliferation of cloud has led companies and resource providers to build large warehouse-sized data centers, in an effort to respond to costumers demand for computing resources. Operating with powerful data centers requires a significant amount of electrical power, which translates into more heat to dissipate, possible thermal imbalances, and increased electricity bills. On the other hand, as data centers grow in size and in complexity, failure events become norms instead of exceptions. However, failures contribute to the energy waste as well, since preceding work of terminated tasks is lost. Therefore, today's cloud data centers are faced with the challenge of reducing operational costs through improved energy utilization while provisioning dependable service to customers. This chapter discusses the causes of power and energy consumption in data centers. The advantages brought by cloud computing on the management of data center resources are discussed, and the state of the art on schemes and strategies to improve power and energy efficiency of computing resources is reviewed. A practical case of energy-efficient and service-level agreement (SLA)-based management of resources, which analyzes and discusses the performance of three state-of-the-art scheduling algorithms to improve energy efficiency, is also included. This chapter concludes with a review of open challenges on strategies to improve power and energy efficiency in data centers.

1. INTRODUCTION

Modern society increasingly depends on large-scale distributed systems, in areas such as business, government, and defence. According to Tanenbaum and Van Steen [1], a distributed system is essentially a collection of independent computers, sharing resources and interacting with each other toward achieving a common goal, via a computer network. In this regarding, a distributed system appears to users as a single coherent system. The shared resources in a distributed system include data, storage capacity, and computational power, and the distributed computing infrastructure can be used to serve diverse objectives, ranging from executing resource-intensive applications (e.g., CPU- and memory-bound applications) to serving scalable Internet applications (e.g., CPU- and network-bound applications). In particular, in the last recent years, large-scale distributed systems have consolidated itself as a very powerful platform to develop and operate a new generation of business, delivered through the Internet. No matter the size or location of the business, the Internet allows almost any business to reach a very large market. And so, distributed systems have become a powerful mean to support the evolution of civilizations, providing services for business, development of research and science, and welfare of populations.

Recently, an increasingly interest in Service-Oriented Architectures (SOAs) [2], which focus predominantly on ways of developing, publishing, and integrating application logic and/or resources as services, and the constantly growing solutions for virtualization [3] have lead to the concept of cloud. Cloud computing [4] represents a new type and specialized distributed computing paradigm, providing better use of distributed resources, while offering dynamic, flexible infrastructures and Quality of Service (QoS) guarantees. In spite of the advantages brought by the cloud computing emerging business model, nowadays Information Technology (IT) industry is adopting cloud to offer users high-available, reliable, scalable, and inexpensive dynamic computing environments. In order to deal with the increasing demand for computing resources by end-users, companies and resource providers are building large warehouse-sized data centers. Furthermore, clusters greater than 10,000 processors [5] have become routine in worldwide laboratories and supercomputer centers, and clusters with dozens and even hundreds of processors are now routine on university campuses [6]. As data centers infrastructures proliferate and grow in size and in complexity, the articulation of provided services and deployed resources imposes new challenges to the management of such computing environments regarding failures and energy consumption. In fact, component failures become norms instead of exceptions in large-scale computing environments, which contribute to the energy waste, since preceding work of terminated tasks is lost. At the same time, the amount of electrical energy consumed by data centers increases as the computing power installed grows. On the other hand, research and development in large-scale distributed systems over the last years had been mostly driven by performance, whereas rises in energy consumption were generally ignored. The result was a steady growing in the performance, driven by more efficient system design and increasing density of the components according to Moore's law [7]. Regrettably, the total power drawn by computing systems has not been following the constant raise in performance per watt ratio [8]. The fact is that the average power

Table 1 Estimated Total Power Consumption by Servers in the World (W/U) [8].

Year	Volume	Mid-range	High-end
2000	183	423	4874
2003	214	522	5815
2004	216	578	6783
2005	222	607	8106

use in different types of servers, as depicted in Table 1, has been increasing every year, and the problem is even worse when one consider large-scale computing infrastructures, such as clusters and data centers. As a consequence, the energy consumption in modern data centers accounts for a considerably large slice of operational expenses. Koomey [9] estimated that the energy consumption in data centers has risen by 56% from 2005 to 2010, and in 2010 accounted to be between 1.1% and 1.5% of the global electricity use. Today, the average data center consumes as much energy as 25,000 households [10], and estimations point out it will continue to grow in the future. Taking into account that from a cloud providers perspective the maximization of the profit is a high priority, the optimization of energy consumption plays a crucial role in the reduction of operational costs.

Tightly coupled with energy consumption, data centers have a large and growing substantial CO_2 footprint. Kaplan study [10] estimates that today's data centers result in more carbon emissions than both Argentina and The Netherlands. In spite of this, the continuous increase in data centers energy consumption and the inefficiency in data centers energy management have now become a major source of concern in a society increasingly dependent on IT. These cost and environmental concerns have already been prompting many "green computing" energy initiatives, with the aim at reducing the carbon footprints. Raised by these environmental concerns, governments worldwide are approving laws to regulate the carbon footprint. For example, the European Union issued a voluntary Code of Conduct in 2007 prescribing energy efficiency best practices [11]. Another example of the efforts taken in this direction is the Green Grid association, a global industry consortium focused on promoting energy efficiency for data centers and minimization of the environmental impact, which has also forwarded recommendations for improved metrics, standards, and technologies. The scope of sustainable computing includes not only computing nodes and their components (e.g., processors, memory, storage devices) but also other kinds of related resources, such as auxiliary equipments for the computing facilities, water used for cooling, and even physical space that all these resources occupy.

This chapter presents an overview about energy-efficient management of resources in cloud data centers, with QoS constraints. Several techniques and solutions to improve the energy efficiency of computing systems are presented, and recent research works that deal with power and energy efficiency are discussed. The reminder of the chapter is organized as follows: Section 1 introduces basic concepts related to the subject; Section 2 identifies and discusses the sources of power and energy consumption of data centers, as well models to represent their behavior; Section 3 introduces cloud computing, a distributed model to build virtual clusters on top of data center infrastructures, and discusses some advantageous techniques inherent to clouds to enforce power and energy efficiency; Section 4 identifies two important factors that constraint the energy efficiency in today's data centers; Section 5 presents a survey about relevant state-of-the-art strategies and schemes to improve energy efficiency in cloud data centers; Section 6 presents a study over three possible alternatives to improve the energy efficiency in cloud data centers based on consolidation of VMs, where physical servers are subject to failure; and Section 7 concludes this chapter and points out some research challenges.

2. ENERGETIC CHARACTERIZATION OF A DATA CENTER

This section presents an overview of power and energy concepts associated with any circuit element, analyses and identifies the sources of power consumption in a data center, and presents the mathematical model for power consumption by physical servers in modern data centers.

2.1 Power and Energy Models

In order to understand and determine the power being absorbed by any circuit element, it is essential to clarify associated terminology. Electric current is the rate of charge flow past a given point in an electric circuit, measured in coulombs/second. The standard metric unit for current is the ampere (A). A current of 1 A corresponds to 1 coulomb of charge passing through a cross section of a circuit wire every 1 s. The electric power associated with a complete electric circuit or a circuit component represents the rate at which work is done, and is measured in Watt (W) or Joules per second. In turn, energy is the total amount of work performed over a period of time and is measured in Watt-hour (Wh) or Joule (J). Since the energy in Wh relates to the power expended for 1 h of time (commonly used in electrical applications), while the energy in Joule (standard unit of energy in electronics and general scientific applications) is the equivalent of 1 W of power radiated or dissipated for 1 s, then an energy expenditure of 1 Wh represents 3600 J. For example, doing 100 J of work in 1 s (using 100 J of energy), the power is 100 W. The concept of work in the context of computers involves activities associated with running applications (e.g., addition, subtraction, memory operations). Mathematical expressions for power and energy are defined in Eqs. (1) and (2), respectively.

$$P = \frac{W}{T}, \tag{1}$$

$$E = P \times T, \tag{2}$$

where P is the electrical power, T is the period of time, W is the total work performed during the period of time T, and E is the energy. The distinction between the concepts of power and energy is crucial, because techniques that reduce power do not necessarily reduce energy. For example, the power consumed by a server can be decreased by lowering the CPU performance [12], with consequent slowdown in execution of applications. In certain cases, the slowdown can be so extreme that despite saving power, the total energy consumed will be similar [13]. In turn, energy consumption can be decreased by using energy-efficient technologies and optimized hardware/ software, and running workloads on as few servers as possible. Idle servers can be switched to low power modes. If reduction of power consumption results in decreased costs of the infrastructure provisioning (e.g., costs of power generators, power distribution equipment, cooling system), reduction in energy consumption means decreasing the electricity costs.

Both power and energy consumption are of concern in server systems. High peak power demands translate into extra costs associated with uninterruptible power supplies (UPSs), power distribution units (PDUs), backup power generators, and power distribution equipment. Also, high peak power demands translate into complex cooling configurations with high cooling capacity. The problem is even more exacerbated with high power densities, posing serious cooling and reliability problems. For example, blade servers, which aim to bring more computational power in less rack space, are hard to cool because of high density of the components per server, large number of servers packed per square meter, and lack of space for the air flow. On the other hand, the energy consumption, which is a function of the power drawn in the data center, is reflected in the electricity bill. The costs of electricity in a data center is dictated by the energy consumed by servers, interconnects, and cooling infrastructure. Thus, while power consumption determines the cost of the infrastructure required to operate the system, the energy consumption accounts for electricity bills. In spite of this, the power and energy requirements of server systems are determinant for fixed and variable operational costs of data centers.

2.2 Monitoring the Power and Energy Consumption

In order to properly manage data center resources to optimize energy consumption, it is essential to monitor the power consumption by different devices of the data center. To measure the unified efficiency of a data center and improve its performance per watt, the Green Grid association introduced the Power Usage Effectiveness (PUE), one of the today's most used metrics to evaluate how much energy is being deployed to produce useful work. The PUE is defined as the ratio of the total power consumption of a facility (data or switching center) to the total power consumption of IT equipment (servers, storage, routers, etc.), as expressed by Eq. (3).

$$PUE = \frac{\text{Total Facility Power}}{\text{IT Equipment Power}} \tag{3}$$

The PUE is a factor greater or equal to 1, and according to US EPA report [14], it amounts to 1.9 on average for current data centers in the world. In other words, this means that the infrastructure overhead (e.g., cooling, power distribution) represents additional 0.9 W for every watt of power consumed in the computing equipment. Another industry standard metric for measuring data center power consumption is Data Center Infrastructure Efficiency (DCIE), which is the reciprocal of PUE, and calculated as indicated by Eq. (4).

$$DCIE = \frac{1}{PUE} \tag{4}$$

Both PUE and DCIE metrics introduced by the Green Grid focused on increasing the energy efficiency of data centers. Other metrics to measure the performance of data centers in terms of energy can be found in Refs. [15, 16].

The energy consumption of computing resources can be determined by energy sensors, such as wattmeters that provide accurate and periodical measurements, or be estimated by energy models. Wattmeters can be external equipment or components packaged in PDUs that log the power consumption of the node each second. More recently, wattmeters appear as well in the Intelligent Platform Management Interface (IPMI) [17] cards embedded in the computers themselves. For example, to measure the power consumed by a server, IPMI framework provides an uniform interface to access the power monitoring sensors available on recent servers, independent of the operating system. Other alternatives use energy models to estimate the energy consumption of the whole server, based on Advanced Configuration and Power Interface (ACPI)-enabled power supplies. Systems for measuring the power consumed by compute clusters have been described in the literature [18].

Wattmeters available in the market vary in terms of connection interface, communication protocols, packaging, measurement precision, and manner they operate. For example, wattmeters can work in a push mode basis, whereas others respond to queries, sending data to an administration network, isolated from the data center main data network, via Ethernet.

2.3 Sources of Power Consumption

The total power consumption of data centers in the USA and EU has been growing rapidly over the last 10 years. In spite of this, power and energy consumption constitute a big concern nowadays for system architectures and engineers, because they represent extra operational cost and imply environmental impact in terms of CO_2 emissions. To take actions to optimize power and energy consumption, it is paramount to understand the sources of such consumption. This section discusses the sources of power consumption at data center, server, and CPU levels.

According to Lannoo *et al.* [19], data centers worldwide are estimated to have consumed 268 TWh in 2012. A relevant fraction of the total energy consumed is devoted to the cooling systems and other overheads, as registered in Table 2.

Table 2 Data Centers Worldwide Power Consumption in 2012 [19].

Server Class	Volume	Mid-range	High-end
Power per server	222 W	607 W	8106 W
Storage		24% of total server power consumption	
Communication		15% of total server power c onsumption	
Infrastructure		PUE = 1.82	
Total energy	219 TWh	15 TWh	34 TWh

It is clear that the scope of sustainable computing includes not only computing nodes and their components (e.g., processors, memory, storage devices) but also other kinds of related resources, such as auxiliary equipments for the computing facilities, water used for cooling, and even physical space that all these resources occupy. Thus, in order to reduce data centers inefficiencies (i.e., high electricity costs, carbon emissions, and server overload and low QoS), current solutions include (i) more efficient cooling, to improve the PUE; (ii) adoption of energy-efficient servers (e.g., voltage and frequency scaling); and (iii) improvement of resources usage. Regarding the IT equipment, the fraction of power consumption for storage and communication equipment represents 24% and 15% of the total IT equipment power consumption, respectively [19]. These values demonstrate that the most significant fraction of overall data centers power consumption is dictated by servers.

In an effort to reduce the power consumption of cooling facilities, different solutions have been proposed [20–22], and some data centers, such as Google [5], have been operating with a PUE of 1.12. Some companies initiatives are designing data centers from the ground up for energy efficiency, by either generating their own renewable energy or drawing power directly from a nearby renewable power plant. For example, in this regarding Apple is building a 20 MW solar array for its data center in Reno, Nevada [23].

Modern data center servers are equipped with processors that are able to operate at multiple frequency and voltage levels. To minimize the energy consumed by computing nodes, the operating systems or resource managers can choose the level that matches the current workload. This approach works at hardware level and is described in Section 2.6.

The third alternative, described above, to improve the energy efficiency in the data center, is carried out at the resource management level. It implies to dynamically monitor resources usage and correct eventual power inefficiencies. By means of workload consolidation and rescheduling requests, active nodes can transit to low power consumption modes. The process of workload consolidation consists in exploiting (and even exploring by means of prediction) fine-grained fluctuations in the application workloads and continuously reallocating/aggregating as much workloads as possible in the least number of physical servers. The final objective is to improve the utilization of computing resources and reduce energy consumption under workload QoS constraints. These techniques are further described in Section 5.

2.4 Distribution of Server Power Consumption per Component

The cost for the energy consumption of a typical computing server during its lifetime surpasses its purchase cost [24]. According to Barroso and Hölzle [25], the approximate distribution of peak power usage by hardware subsystem in one of Google's data centers is the one illustrated in Fig. 1.

As observed, the main part of power consumption in a server is dictated by the CPU, followed by the memory and losses due to the power supply inefficiency. However, the power required by each component depends on the workload characteristics [16]. Specific techniques focusing on reducing the power consumption of each server component are out of the scope of this chapter. More information about this subject can be found in Ref. [26].

2.5 Modeling Power Consumption by Servers

Based on Fig. 1 that points the CPU as the component consuming more power in a server, it is important to understand how the power consumption by a host evolves regarding the CPU load. According to several studies and experiments (e.g., Refs. [27, 28]), the power consumption by a server can be assumed to grow linearly with the growth of the CPU utilization from the power corresponding to the idle state to the power consumed when the server is fully utilized.

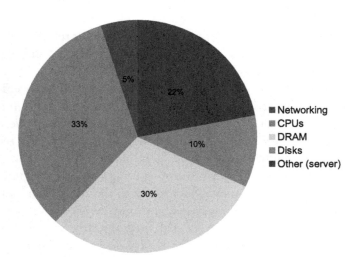

Figure 1 Approximate distribution of peak power usage by hardware subsystem in one of Google's data centers [25].

This behavior is captured by Eq. (5).

$$P(\text{CPU}) = P_{\text{idle}} + (P_{\text{max}} - P_{\text{idle}}) \times \text{CPU} \qquad (5)$$

where P is the estimated power consumption by a physical server, P_{idle} represents the power consumption in idle state (CPU $= 0$), P_{max} is the power consumed by the server when it is fully utilized (CPU $= 1$), and CPU represents the current CPU utilization. The factor $(P_{\text{max}} - P_{\text{idle}})$ is the dynamic power consumed by a physical server. While this equation allows to determine the power consumption by a physical server with an error below 10%, Fan *et al.* [29] have derived more accurate nonlinear relations, described in Eq. (6).

$$P(\text{CPU}) = P_{\text{idle}} + (P_{\text{max}} - P_{\text{idle}}) \times (2 \times \text{CPU} - \text{CPU}^r) \qquad (6)$$

where r is a calibration parameter that minimizes the square error and has to be calibrated experimentally for each class of machines of interest. Most of the research works present in literature use Eq. (5) in analytical and simulation experiments, since refinements of Eq. (6) have little practical utility.

Based on Eq. (5), Xu and Fortes [30] introduced the concept of power efficiency E_{P_i}, for a physical node i. In this regarding, E_{P_i} is given by Eq. (7), and it reflects how much useful work is produced in node i for a given power consumption

$$E_{P_i} = \frac{\text{CPU}_i \times P_{\text{max}}}{P_{\text{idle}} + (P_{\text{max}} - P_{\text{idle}}) \times \text{CPU}_i}. \qquad (7)$$

The power efficiency of a PM i increases monotonically with the workload, reaching 1 at 100% of CPU usage.

2.6 Modeling Power Consumption in CPU

Modern processors are built over complementary metal–oxide semiconductor (CMOS) technology. The main power consumption in CMOS circuits comprises two forms of power, dynamic power consumption P_{dynamic} and static power consumption P_{static}, as indicated by Eq. (8). The static power consumption, also known as idle power or leakage, is the dominant source of power consumption in circuits [13], persisting whether a computer is active or idle. Leakage power consumption is caused by parasitic current that flows through transistors, due to their construction, even when the transistor is switched off. In addition to consuming static power, computer components also consume dynamic power due to circuit activity such as the changes of inputs in an adder or values in a register. Dynamic power can be approximately determined by Eq. (9).

$$P = P_{static} + P_{dynamic}, \tag{8}$$

$$P_{dynamic} = aCV^2f, \tag{9}$$

where a is the switching activity factor that relates to how many transitions occur between digital states (i.e., 0 to 1 or vice versa) in the processor, C is the total capacitance load, V is the supply voltage, and f is the clock frequency of the processor. According to Eq. (9), there are four ways to reduce the dynamic power consumption. However, changing the switching activity a and the total capacitance load C implies to change low-level system design parameters such as transistor sizes and wire lengths. On the other hand, combined reduction of the supply voltage V and the clock frequency f is called dynamic voltage and frequency scaling (DVFS). Because dynamic power is quadratic in voltage and linear in frequency, adjusting the voltage and frequency of the CPU reduces the power dissipation cubically. In this regarding, when a server is not fully utilized, DVFS can be used to downscale the CPU performance and, consequently, reduce the power consumption. DVFS is supported by most modern CPUs, such as the Intel p7.

Several approaches in the literature tried DVFS technology to reduce energy consumption. A recent approach proposed by Ding et al. [31] focused on dynamic scheduling of workloads to achieve energy efficiency and satisfy deadline constraints. The considered scenario comprehends a cloud infrastructure comprising a set of heterogeneous physical servers. Authors explore the idea that there exists optimal frequency for a physical server to process certain workloads. The optimal frequency defines the optimal performance–power ratio to weight the heterogeneities of the physical servers. The physical server with highest optimal performance–power ratio is selected to process the workloads first, in order to save energy, unless it does not have enough computation resources. Authors proposed the EEVS scheduling algorithm, which supports DVFS technology to improve energy efficiency. The scheduling is divided into some equivalent periods, in each of which workloads are allocated to proper physical servers and each active core operates on the optimal frequency. After each period, the cloud is reconfigured to consolidate the computation resources of the physical servers to further reduce the energy consumption. This particular work ignores the performance and power penalties of status transitions of processor and workload migrations, limiting its deployment in real cloud environments.

Despite DVFS can result in substantial energy savings, the technique has limitations such as difficulty to deal with the complexity of modern CPU architectures (i.e., pipelining, hyperthreading, etc.), which makes nontrivial to determine the required CPU clock frequency to meet application's performance requirements. Also, power consumption by a CPU in real systems may not be quadratic to its supply voltage, due to nondeterminism which breaks any strict relationship between clock frequency, execution time, and power consumption.

2.7 Minimize Power Consumption and Improve Energy Efficiency

Although DVFS promises to reduce hosts' power consumption by dynamically changing the voltage and frequency of a CPU according to its load, the adoption of multi-core CPUs along with the increasing use of virtualization resulted in the growing amount of memory in servers, which have narrower dynamic power range [29]. Additionally, even if a server is completely idle (i.e., CPU has no activity), it still consumes 50–70% of the power it would consume when fully utilized, due to I/O operations, memory, etc., as demonstrated by several studies (e.g., Refs. [29, 32–34]). Even reducing the power consumption up to 70% by applying very low CPU activity modes [35], other system components have narrower dynamic power ranges (e.g., less than 50% for DRAM, 25% for disk drives, and 15% for network switches). Regarding power supplies which transform alternating current (AC) into direct current (DC), they achieve the highest efficiency at loads within the range of 40–60%, and almost maintaining it till 100%, as described in Fig. 2. Considering that most data centers normally create a load of 11–50% [36], and that DVFS optimization is limited to CPUs, alternatives for reducing infrastructure energy consumption, such as consolidation of workloads via scheduling policies to reduce the number of active servers and hence the overall power consumption, are more attractive.

In the context of exploring the transition of servers to/from low-power modes, Meisner *et al.* [38] proposed PowerNap which allows an entire server to rapidly transit between a high-performance active state to a near-zero-power idle state. In this sense, PowerNap energy-conservation approach aims at minimizing idle power and transition time, instead of requiring fine-grained power–performance states and complex load-proportional operation from server's components. Authors demonstrate that it is possible to substantially reduce the power consumption from 270 W in idle state to only 10.4 W in nap mode, in about 0.3 ms, for a typical blade server.

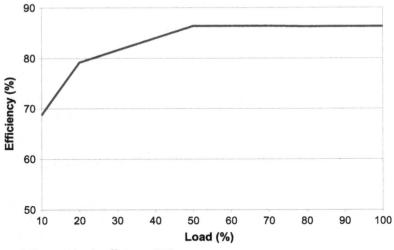

Figure 2 Power supply efficiency [37].

Comparing PowerNap with DVFS approach, the former shows to be more practical and efficient, in the sense that it requires system components to support only two operating modes, and provides greater energy efficiency and lower response time. According to PowerNap authors, the approach is suitable for real-world systems because the delay of transitions from and to the active state is less than the period of inactivity of most real-world systems, and the saved power is more than the power required to reinitialize the component from nap state.

3. CLOUD COMPUTING ENVIRONMENTS

In general, all main trends in IT are based on large and powerful computing data centers. Cloud computing is the most prominent example. This section discusses this recent computation model, in terms of advantages, limitations, and technologies associated to empower computational resource management.

3.1 Introduction

Cloud computing [4] aim is to make a better use of distributed resources, which the massive computers constitute, while offering dynamic flexible infrastructures and QoS-guaranteed services. Key concepts distinguishing cloud computing from other paradigms include on-demand, high-availability, and high-scalability access to large-scale distributed resources, with QoS-driven management [39]. From a hardware point of view, users have the illusion of infinite computing resources that are available on demand. By using cloud-based systems, costumers can have easy access to large distributed infrastructures, with the ability to scale up and down the assigned computing resources according to their applications' needs. Such ability is known as elasticity and refers to the capability of the underlying infrastructure to adapt to changing (e.g., amount and size of data supported by an application, number of concurrent users), by applying vertical and horizontal elasticity of resources. While vertical elasticity relates to the fraction of hardware resources assigned to workloads, horizontal elasticity refers to the amount of hardware instances necessary to satisfy the workload variability. Hence, the capability of rapid elasticity of clouds represents an enormous advantage compared to other computing paradigms since cloud users can concentrate specifically on developing their work, without concerning about over- or under-provisioning of resources. Another important feature of cloud is that organizations pay for computing resources on a consumption basis, very much like the utility companies charge for basic utilities such as electricity, potentially leading to cost savings.

Discussed key advantages make clouds to be progressively adopted in different scenarios, such as business applications, social networking, scientific computation, and data analysis experiments. Proliferation of cloud computing and consequently the establishment of new large-scale data centers all over the world containing massive number of interconnected compute nodes brought new challenges to the management of these computing environments regarding energy consumption and satisfaction of applications' QoS constraints.

3.2 Service Types and Deployment Modes

According to Schubert *et al.* [40], clouds can be classified according to their service types and deployment modes.

- **Infrastructure as a Service (IaaS)**: It provides resources that have been virtualized (see Section 3.3) as services to the user, such as virtual computers, meeting various computing needs, and even virtual clusters. Cloud infrastructures apply resource elasticity and workload consolidation techniques to rapidly adapt to environment changes (i.e., variation in computational capacity needs, physical nodes availability, etc.), and provide on-demand access to computational resources in a pay-as-you-go basis. Common operation in clouds implies that provisioning of services is specified and regulated by ensuring the desired QoS, and requests for instantiation of services may be accepted or rejected based on current and predicted infrastructure load and capacity. Among public IaaS providers, Amazon EC2[1] offers access to EC2 instances (i.e., virtual execution environments—refer to Section 3.3 for more details) which looks much like physical hardware, letting users control the software from the kernel upward. However, this approach has inherent difficulty supporting automatic scalability and failover, since traditional methods, such as replication, are highly application dependent.

- **Platform as a Service (PaaS)**: It provides developers with an environment upon which applications and services can be developed and hosted. Typically, users control the behavior and execution of a server hosting engine through dedicated Application Programming Interface (API) deployed by cloud providers. In this way, developers can concentrate on creating services instead of worrying about building an environment for running them.
- **Software as a Service (SaaS)**: It provides access to software packages/services using the cloud infrastructure or platform, but keeping users unaware of underlying infrastructure. SaaS allows end-users to avoid rigid license agreements or infrastructure costs.

Furthermore, clouds can be classified based on the business model of the provider as:

- **Private Clouds**: Typically owned by the respective organization, services deployed can be accessed and managed exclusively by people of the organization, thus not being directly exposed to the customer.
- **Public Clouds**: Organizations may offer their own cloud services to clients or other service providers, thus being available to general public. In this context, enterprises can utilize such cloud providers' services to avoid build up their own infrastructure with consequent cost reduction.
- **Hybrid Clouds**: It consists of mixing private and public cloud infrastructures and services. This model allows enterprises to keep control over sensitive data by employing local private clouds, and to outsource parts of their infrastructure to public clouds so as to achieve a maximum of cost reduction.

---

[1] Amazon EC2.http://aws.amazon.com/ec2/.

Additional classifications can eventually be found in literature, or at different granularity. However, the classification here presented identifies the main types of clouds currently in use.

3.3 Virtualization Concept and Features

Virtualization [3] is the main technology in cloud computing to deploy IaaS. It promises a reduction in cost and complexity through the abstraction of computing resources such that a single physical server is able to support a large number of disparate applications running simultaneously. In turn, each application runs in a logical container—the virtual machine (VM). In this sense, virtualization enables a reduction in the number of physical components, meaning fewer assets to configure and monitor, and consequent reduction in complexity and cost of managing cloud infrastructures.

The concept of virtualization was first introduced with the IBM mainframe systems in the 1960s, to refer to a VM [41]. A virtualization system is an architecture able to separate and isolate an operating system from the underlying hardware resources and lower-level functionalities (e.g., servers, network links, and host bus adapters). Server virtualization [42] is performed on a given hardware platform by introducing a thin hypervisor layer, sometimes called virtual machine monitor (VMM), which sits on top of the physical hardware resources and creates a simulated computer environment (i.e., a VM).

Hypervisors can be implemented in server firmware or in software [43]. Virtualization splits a single physical computer into multiple logical computers, or VMs, each accommodating its own guest software, as described in Fig. 3. Then, the hypervisor multiplexes and arbitrates access to resources of the host platform so that they can be shared efficiently among multiple VMs. The guests run just as if they were installed on a stand-alone hardware platform. The number of VMs per PM is limited by the host hardware capability, such as core number, CPU power, memory resources, and can be defined as well by the software provider. According to the type of virtualization, there is no requirement for a guest operating system (GOS) to be the same as the physical host one. Guests are able to access specific hardware components, such as hard disk drives and network interfaces.

The recent resurgence of virtualization relates to the possibility to deal with issues in the data center, such as variance in demanding requirements of computing resources, power inefficiency due to hardware underutilization, high system administration costs, and reliability and high availability.

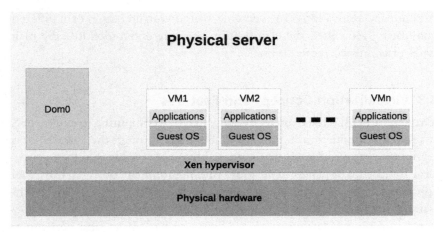

Figure 3 Virtualization concept.

Current virtualization technologies provide powerful mechanisms to tackle these issues, namely VMs resizing, migration, and checkpointing. VM resizing mechanism permits the adjustment of resources allocation, by adding additional CPUs, network interfaces, or memory. In turn, VM migration [44] mechanism gives the ability to easily move VMs from one physical node to another one using live or stop and copy migration alternatives. Both VM resizing and migration techniques are the necessary conditions to achieve elasticity and the appearance of infinite capacity available on demand. At the same time, it benefits cost reduction by server consolidation, since the number of hosting servers can be minimized based on better utilization of the expensive hardware resources. The checkpoint mechanism works asynchronously by replicate/checkpointing the primary VM memory and disk to a backup one, in a very high frequency [45–47], thus providing fault tolerance. Administrators can use the previously mentioned tools to adequately manage the computing environment.

Therefore, through resource elasticity and server consolidation techniques, cloud infrastructures are able to rapidly adapt to energy inefficiencies and failures in physical nodes, providing high-availability access to computational resources on demand. Thus, VMs resizing and migration have emerged nowadays as a promising technique to be utilized by resource management algorithms since it can cope with resources allocation problems in modern data centers. For example, based on this techniques, Lagar-Cavilla et al. [48] developed SnowFlock system, a VM fork approach that rapidly clones a VM into several replicas running on different PMs. Similar to

Ref. [49], it uses fast VM instantiation by initially copying and transmitting only the critical metadata (authors called it VM descriptor) necessary to start execution on remote host. Memory-on-demand mechanism fetches portions of VM state, from an immutable copy of the VM's memory from the time of cloning, over the network as it is accessed. Cooperatively, avoidance heuristics algorithms try to reduce superfluous memory transfers that will be immediately overwritten, and multicast replies to memory page requests. These key techniques enable SnowFlock to introduce little runtime overhead and small consumption of I/O resources, leading to good scalability. VMs disks are implemented with blocktap driver. SnowFlock represents an example how modern data centers can leverage virtualization to provide excess load handling, opportunistic job placement or even parallel computing, as well as to provide fault tolerance due to replication of VM instances.

Nowadays, there are several system-level virtualization solutions offered by some commercial companies and open-source projects. Some of the major virtualization software are Xen,[2] depicted in Fig. 3, KVM,[3] and VMware.[4]

In the case of Xen, the technology currently supports various scheduling algorithms to schedule domains (i.e., VMs); some of them were extensively analyzed to determine their performance considering different scheduling parameters [50–52]. Xen CPU credit scheduler[5] is the default scheduler in Xen since version 3.0 and is designed to ensure that each VM and/or a virtual CPU (VCPU) gets a fair share of the physical CPU resource. In the credit scheduler, each domain is given the opportunity to configure CPU affinity and priority, by means of pinning VCPUs to specific cores and defining cap and weight scheduler parameters, respectively. The cap scheduler parameter specifies the maximum percentage of CPU resources that a VM can get, while weight parameter determines the number of credits associated with the VM, which controls the priority in the execution of VMs. According to those weights (e.g., number of credits), the scheduler assigns physical CPU time allocation among all the VMs. For example, Xen credit scheduler can allow execution of two tasks, one VM receiving 30% of total CPU capacity and the other the 70% remaining. In this regarding, there are examples of strategies setting the cap parameter to adjust

[2] http://www.xensource.com/.
[3] http://www.linux-kvm.org/.
[4] http://www.vmware.com/.
[5] http://wiki.xen.org/wiki/Credit_Scheduler.

dynamically resource allocation based on task usage, and to make sure co-located tasks cannot consume more resources than those allocated to them [53].

3.4 Support to Workload QoS Requirements

The current service-oriented economy trend [54], supported by distributed computing paradigms such as cloud computing, has led to an increased concern about aspects regarding the quality and reliability of the services offered. As consumers move toward adopting such SOAs, the exploitation of service-oriented technologies is followed by a growing interest of QoS mechanisms [55]. However, providing QoS guarantees to cloud users is a very difficult and complex task, because the demands of the service consumers vary significantly, and initially assigned computing resources to applications tend not to perform as expected as they are being shared among different types of workloads. In this regarding, services to costumers are often provided within the context of SLA, which ensures the desired QoS. SLAs can be defined as a static mechanism of agreements between users and service providers, in which all expectations and obligations of a service are explicitly defined [56]. Its main objective is to serve as the foundation for the expected level of service between the consumer and the provider, translating users' expectations in terms of QoS into infrastructure-level requirements. This way, users define their QoS requirements in high-level SLA metrics, which in turn correspond to low-level resource metrics used by cloud management system to enforce service utilization policies and usage commitments. The QoS attributes are generally part of an SLA, such as response time, throughput, and maximum execution time to accomplish a task, and need to be closely monitored so management system can guarantee QoS on each application execution [57]. An SLA violation results in a penalty incurred by the cloud provider.

Virtualization technology is useful for cloud providers, in assisting the management of IaaS to fulfill users' applications QoS requirements. The correct and effective use of the features provided by virtualization, such as migration and Xen credit scheduler parameters, allows cloud providers to cope with important current data centers issues, such as reduction in energy costs. Since a single PM is able to support a large number of disparate applications running simultaneously within logical containers (i.e., VMs), by means of adequate scheduling of VMs workloads can be consolidated on fewer servers (i.e., unneeded servers can be hibernated or used to accommodate more load), resulting in reduction of operational and energy costs.

4. CONSTRAINTS TO ENERGY EFFICIENCY IN CLOUD DATA CENTERS

Cloud providers are faced with new challenges such as reduction of energy consumption and guaranteeing SLAs, the key element to support and empower QoS in these environments. Despite the advantages from virtualization to manage resources in a flexible and efficient way, the growing complexity of data centers allied with the diversity of applications submitted results in more nodes to manage and greater system management difficulty as well. This section addresses three important factors that contribute to constrain the energy efficiency in cloud data centers.

4.1 Workloads Characteristics

Today's clouds are being adopted in different scenarios, such as business applications, social networking, scientific computation, and data analysis experiments [40, 58, 59]. Therefore, current clouds host a wider range of applications with diverse resource needs and QoS requirements. In this regarding, efficient provisioning of resources became a challenging problem in cloud computing environments due to its dynamic nature and the richness of application workload characteristics. Consolidation of VMs is a very common technique to improve resource utilization (i.e., cloud providers try to schedule multiple VMs onto fewer servers), with the objective to reduce power consumption, and consequently operational costs. However, optimizing energy efficiency through consolidation of VMs is challenging since it can lead to degradation in applications' performance. The counter effect of consolidation can be even more severe depending on the type of workload a VM is running, which makes previous scheduling conditions outdated. For example, in the case of web applications, in which workloads demand is bursty, outdated schedules can result in situations of scarcity of resources, with consequent penalty in QoS of applications perceived by end–users. On the other hand, over-provisioning of resources to applications is not a solution in the sense that it will result in waste of energy. This problem is even more serious since current cloud data centers either do not offer any performance guarantee or prefer static VM allocation over dynamic. For example, Amazon EC2 offers only guarantees on availability of resources, not on performance of applications running within VMs [60]. In this regard, it is imperative to consider these limitations in the scheduling of applications in the cloud.

Existing research has addressed many aspects of this problem including, for example, efficient and on-demand resource provisioning in response to dynamic workload changes [60–63], and platform heterogeneity-aware mapping of workloads [64, 65]. However, in order to improve resource management and operational conditions while QoS guarantees are maintained, it is critical to understand the characteristics and patterns of workloads within a cloud computing environment. Fortunately, some recent notable efforts have provided a comprehensive in-depth statistical analysis of the characteristics of workload diversity within a large-scale production cloud [66–68]. On the basis of these studies is the second version of the Google Cloud tracelogs [69], which contains over 12,000 servers, 25 million tasks, and 930 users over the period of 29 days. These studies yielded significant data on the characteristics of submitted workloads and the management of cluster machines. Their contribution enables further work on important issues, such as resource optimization, energy efficiency improvements, and failure correlation. The analysis performed clarified that approximately 75% of jobs only run one task and most of the jobs have less than 28 tasks that determine the overall system throughput. The average length of a job is 3 min, and the majority of jobs run in less than 15 min, despite the fact that there are a small number of jobs that run longer than 300 min. Moreover, task length follows a lognormal distribution, with most of the tasks requiring a short amount of time. This same distribution applies to CPU usage, which varies from near 0% to approximately 25%, indicating that a high proportion of tasks consume resources at lower rates. For the same reason, a lognormal distribution can be applied to describe the number of tasks per job. Depending on the cluster or day observed, job inter-arrival times follow distributions such as lognormal, gamma, Weibull, or even exponential, with a mean time of 4 s. Most of the tasks use less than 2.5% of the node's RAM.

Derived from these key findings about workloads characteristics executing in large-scale data centers, some models were implemented by extending the capabilities of the CloudSim framework [70], and further validated through empirical comparison and statistical hypothesis tests. In particular, the implementation in a simulation framework constitutes an important contribution because discrete-event simulation is usually chosen as a first step toward creating cloud computing environments due to the ability to guarantee repeatable conditions for such a set of experiments. Moreover, the authors of these studies also pointed out several examples of their work's practical applicability in the domain of resource management and energy efficiency.

4.2 Failures and Unavailability Characteristics

Cloud systems focus on being scalable, high–available, cost–efficient, and easy access to resources on demand. As cloud systems are progressively being adopted in different scenarios, such as in scientific computing [59], availability of a massive number of computers is required for performing large-scale experiments. Nevertheless, some applications take a very long time to execute, even on today's fastest computer systems, which makes significant the probability of failure during execution, as well as the cost of such a failure in terms of energy and SLA violations. Moreover, cloud data centers become more prone to failure as infrastructures continue to grow in size and complexity, involving thousands or even millions of processing, storage, and networking elements. A study conducted by LANL estimated the mean time between failures (MTBF) to be 1.25 h on a petaflop system, extrapolating from current system performance [71]. Based on the same study, an application running on all the nodes of a LANL system can be interrupted and forced into recovery more than two times per day. According to Schroeder and Gibson [72], three of the most difficult and growing problems affecting large-scale computing infrastructures are avoiding, coping, and recovering from failures. In fact, in large-scale networked computing systems, computer component failures become norms instead of exceptions, making it more difficult for applications to make forward progress. An important remark is that the failure rate of a system grows in proportion to the number of processor chips in the system, and technology changes over time have little impact on systems reliability. The consequences can be even more negative to currently executing applications if one consider that several of these nodes are frequently submitted to maintenance operations. The QoS perceived by consumers will be disastrously affected if measures to handle failure are not considered. Furthermore, failures result in a loss of energy because the computation is lost and needs to be repeated after recovery. A recent work by Garraghan et al. [67] has shown the impact of failures in a system in terms of energy costs, which can translate to almost 10% in the context of the total energy consumption for the entire data center. To this end, it is evident that the adoption of fault tolerance mechanisms is crucial in reducing energy consumption and providing dependable systems.

The root cause of the failures occurring in a system can be classified into human, environment, network, software, hardware, and unknown. Failures in hardware represent the largest source of malfunction, accounting for more than 50% of all failures, while failures due to software place around 20% of all failures [72]. Reliability and availability [73] are two measures applying how good the system is and how frequently it goes down. Reliability can be defined as the probability of no failure within a given operating period. In turn, availability is the probability that an item is up at any point in time. Both reliability and availability are used extensively as a measure of performance and to compare the effectiveness of various fault-tolerant methods. Availability of a system can be calculated, in a simple manner, based on uptime and downtime of the system. In this sense, availability is known to be a good metric to measure the beneficial effects of repair on a system. Rapid repair of failures in redundant systems greatly increases both the reliability and availability of such systems. Reliability can be measured by the MTBF, which in turn is estimated by the manufacturer of the component. In turn, mean time to repair (MTTR) is the time taken to repair a failed component, and it expresses responsiveness of the whole system. Availability is often measured by dividing the time the service is available by the total time, as expressed by Eq. (10) [74].

$$\text{Availability} = \frac{\text{MTBF}}{\text{MTBF} + \text{MTTR}} \tag{10}$$

Typically, the availability of a system is specified in nines notation. An availability of 99.9% (3-nines) refers to a downtime of 8.76 h per year. The target availability in most telecommunication devices and many modern data centers is of 5-nines that corresponds to 5 min per year. Both reliability and availability must be considered in order to measure systems performance. Both reliability and availability are attributes of dependability, which refers to the ability of a system to avoid service failures that are more frequent and more severe than is acceptable.

Over the last years, several techniques have been developed to improve systems availability. Avizienis *et al.* [75] grouped these techniques into four major categories, namely: (i) fault prevention, which aims at preventing the occurrence or introduction of faults in development process of systems (both in software and hardware); (ii) fault removal, which tries to reduce the number and severity of faults; (iii) fault tolerance, to avoid service failures in the presence of faults; and (iv) fault forecasting, to determine the future incidence of faults, and related consequences. Recently, fault tolerance and fault forecasting techniques have been used together in an effort to empower systems the ability and confidence in that they are able to deliver a service that can be trusted. Dabrowski [76] defined fault tolerance as "the ability to ensure continuity of service in the presence of faults, or events that cause a system to operate erroneously." Indeed, fault tolerance mechanisms are aimed at failure avoidance, thus giving operational continuity to the system when one of the resources breaks down. Such technique is called failover. For that, fault tolerance mechanisms provide ways to carry out error detection and system recovery. According to Egwutuoha *et al.* [77], the major used fault tolerance techniques are redundancy, migration, failure semantics, failure masking, and recovery. A full revision about fault tolerance and fault-tolerant systems is out of the scope of this chapter. Further information about this subject can be found in Refs. [77, 78]. In short, redundancy, by hardware or software, and failure masking imply replication, which not only increases hardware/software costs and runtime costs (i.e., more energy is spent) but also makes it more difficult to manage. In turn, failure semantics relies on the assumption of the designer to be able to predict failures and specify the most appropriated action to be taken when a failure scenario is detected. However, this assumption is surrealist due to complexity of today's computing environments such as cloud computing. Recovery implies recovering from an error so a system failure can be avoided. Nevertheless, not all states can be recovered, and applications may have to be restarted from scratch. Unlike reactive schemes, which means that faults are dealt after they take place, fault recovery commonly relies on checkpoint/restart mechanisms. Remus system [45] implemented in Xen is an example of reactive fault tolerance. The system continuously copies memory pages, from the VM's private memory address space (i.e., a technique known as copy-on-write (CoW) [79]), to asynchronously replicate/checkpoint the primary VM memory to the backup VM in a very high frequency. The same applies to disk, and the backup VM maintains a precopied disk image file to receive the updates. Unfortunately, these precopies operations are very time consuming and often prove counter-effective [71]. The operation of checkpointing a process or a whole VM can incur significant overhead in a large-scale system. For example, the LANL study [71] estimated that the checkpoint overhead prolongs a 100-h job (in the absence of failure) by an additional 151 h in petaflop systems. In order to circumvent the problem of excessive overhead caused by traditional fault tolerance techniques and additional energy consumption, the trend on system dependability has recently shifted onto failure prediction and autonomic management technologies. Modern hardware devices are designed to support various features and sensors that can monitor the degradation of an attribute over time and allow early failure detection [80, 81]. Server failures can often be anticipated by detecting deteriorating health status, based on data collected using monitoring of fans, temperatures, and disk error logs. At the same time, several statistical and machine learning-based prediction techniques have been presented with up to a considerable accuracy (e.g., Refs. [82–84]). As a result, several research works proposed to estimate the reliability of a server by forecasting when the next failure will occur in that server. Some of the resulting failure prediction tools can be found in Refs. [82–86].

Proactive fault tolerance (PFT) relies on the assumption that node failures can be anticipated based on monitored health status. The technique can be combined with reactive fault tolerance schemes so as to adapt and reduce the checkpoint frequencies, and hence optimizing the trade-off between overhead cost and restart cost. Engelmann *et al.* [87] defined PFT as "(…) a concept that prevents compute node failures from impacting running parallel applications by preemptively migrating application parts away from nodes that are about to fail." Hence, PFT schemes do not need to implement redundancy nor replication. This property represents an advantage to reduce energy consumption and optimize resources utilization in cloud data centers because redundancy means extra power consumption. Moreover, relying on modern VM techniques, such as their ability to reconfiguration through VM migration, systems administrators are allowed to easily schedule plans of maintenance, reconfiguration, and upgrade in the data center infrastructures without interrupting any hosted services. Applications running in nodes that were incorrectly foreseen as reliable will have to restart from scratch in case of node failures. However, in such cases the overhead of restarting a VM is negligible. Lagar-Cavilla *et al.* [48] demonstrated that it is possible to initiate multiple VM instances in less than 1 s, assuming that the VM images and data are stored in a network attached storage. In this sense, the construction of adaptive computing infrastructures with the support of VM technology, which adjust their configuration dynamically at runtime according to components' availability, represents an opportunity to efficiently utilize system resources and tackle energy waste due to failure events.

In order to mask the effect of failures in physical servers and hence to reduce energy waste, it is preponderant to understand failures characteristics. Schroeder and Gibson have conducted a study to characterize the MTBF in a large-scale data center [72]. The key findings show that a Weibull distribution, with a shape parameter of 0.8, approximates well the MTBF for individual nodes, as well as for the entire system. In the case of MTTR, the periods are well modeled by a lognormal distribution, which length depends on the cause of failure.

4.3 Performance Interference Characteristics

A third aspect which constraints the energy efficiency and QoS deviations in a cloud data center relates to technological limitations. Guaranteeing the expected QoS for applications, and reducing the energy consumption at the same time, is not trivial due to QoS deviations. In fact, modern virtualization technologies do not guarantee effective performance isolation between VMs, meaning that the performance of applications can change due to the existence of other co-resident VMs sharing the underlying hardware resources (e.g., Refs. [88–91]). As a consequence, a task running in the same VM on the same hardware but at different times will perform disparately based on the work performed by other VMs on that physical host. This phenomenon has been studied and is known as performance interference. Therefore, optimizing energy efficiency through consolidation of VMs can induce additional degradation in applications' performance, since applications are contending for shared resources. The counter effect of consolidation can be even more severe depending on the type of workload a VM is running. For example, in the case of batch jobs, consolidation can produce severe slowdown due to contention in on-chip resources, meaning that workloads take much longer to complete under consolidation than without it, despite the advantages in energy savings [92].

Dealing with performance deviations is a complex task because some resources are very hard to isolate, such as on-chip shared resources (e.g., cache memory, memory buses), and because of bursty oscillation in demand of resources (e.g., I/O network). Research literature on this subject analyzes diverse causes of performance degradation due to contention in shared resources [91, 93, 94]. Sources of performance interference include contention in Last-Level Cache (LLC) and I/O network. Current multi-core processors are equipped with a hierarchy of caches. Commonly, a CPU chip has one or more cache levels L1/L2 dedicated to each of its cores, and a single shared LLC L3, such as the Intel Nehalem Core i7. For most processors, data is inserted and evicted from the cache hierarchy at the granularity of a cache line of 64 bytes of size [95]. If the co-hosted applications total working set size exceeds the PMs CPU LLC size, then applications will definitely degrade in performance [96]. Because presently it is not possible to measure the cache usage directly, the hypervisor can measure the impact of co-hosted VMs in one another's performances by tracking the misses per instruction, collected through performance counters [97]. Network interface cards (NICs) can also affect the performance of co-hosted VMs. For security reasons, Xen hypervisor imposes that VMs perform I/O operations through special device located at privileged domain Dom0. Each VM uses a virtual NIC (VNIC) that allows exchanging data with Dom0, by sharing memory pages. The notification of VMs is done based on virtual interrupts. High number of received/transmitted packets to/from VMs, as well as small fragmented packets, generates high rate of hypervisor interrupts and data multiplexing to the corresponding VMs. High rate of multiplexing/demultiplexing of packets among Dom0 and VMs creates a communication bottleneck and causes interferences among VMs, such as packet loss for high latency, fragmentations, and increased data copying overhead [98].

Notable research work has contributed with models to capture performance interference effects among consolidated workloads. The extent of degradation clearly depends on the combination of applications that are co-hosted, shared resources, and even the kind of virtualization mode. Therefore, determining the contribution of all these parameters to the interference level on application QoS is critical. Somani and Chaudhary [90] conducted research by running combined benchmarks to give an insight about the isolation properties of Xen hypervisor relating to CPU, network bandwidth, and disk I/O speed resources. The results showed that Xen credit scheduler behaves well when running two nonsimilar resource-intensive applications. Specifically, it provides good isolation when running high-throughput and nonreal-time applications, but it becomes difficult to predict the performance and the time guarantees when running soft real-time applications. These insights contribute to place applications in the data center in order to get the maximum isolation and fairness. In a later work [99], the authors studied the effect on performance interference from different CPU scheduling configurations in Xen hypervisor, for combined I/O- and CPU-intensive applications. Findings reveal that Xen credit scheduler is relatively good in the case of dedicated CPUs, which can be achieved by means of assigning the virtual CPUs (VCPUs) to physical cores (called core pinning or core affinity). For nonpinning configurations, credit scheduler shows much higher time to complete the tests, as compared to the pinned case. This is due to the opposite performance impact of unwanted load balancing which results in more VCPU switches and cache warming as co-hosted domains are doing heavy disk I/O. In running CPU-intensive applications, credit scheduler proved itself good and even showed better performance than Simple Earliest Deadline First (SEDF) scheduler when running heavy I/O (net and disk). For interactive applications, such as games and office applications, SEDF scheduler showed good average latency, with minimum latency similar to credit scheduler. The credit scheduler can, in some cases, result in higher peaks of latency for CPU–intensive applications because of its less prompt service behavior.

Koh *et al.* [100] studied the effects of performance interference by co-locating combined sample applications chosen from a set of benchmark and real-world workloads. Those workloads comprehended different characteristics in terms of CPU, cache, and I/O utilization. Then, authors collected the performance metrics and runtime characteristics for subsequent analysis and identified applications combinations that generate certain types of performance interference. For example, some findings indicate that there is a significant degree of performance interference between I/O–intensive applications, and memory-intensive applications can suffer a performance degradation from 20% to 30%. On the other hand, CPU–intensive programs are able to achieve better performance with I/O–intensive applications. Xen was the virtualization technology used. Authors also proposed a weighted mean method to predict the expected performance of applications based on their workload characteristics, with a 5% average error. A later work [98] extended the study to performance interference among VMs, but in the context of network I/O workloads that are either CPU bound or net-work bound. Authors pointed out interesting findings about combination of types of workloads, namely: (i) network–intensive workloads can lead to high overheads due to extensive context switches and events in Dom0; (ii) CPU–intensive workloads can incur high CPU contention due to the demand for fast memory pages exchanges in I/O channel; and (iii) CPU–intensive workloads combined with network–intensive workloads incur the least resource contention, delivering higher aggregate performance. Considering that experiments were taken using Xen hypervisor, authors concluded additionally that the efficiency of Dom0 highly affects the perfor-mance of multiple VMs.

Lim *et al.* [101] proposed D-factor model, which models the problem of performance interference within a physical server based on a collection of multiple-resource queues for which contention creates nonlinear dilation factors of jobs. Authors tested CPU- and disk I/O-bound applications. Key findings revealed (i) multiple-resource contention creates nonlinear dila-tion factors of jobs; (ii) linear relationship between total completion time and individual completion times for the same type of jobs; and (iii) co-location of workloads utilizing different system resources leads to best efficiency. Authors claim that proposed D-factor model is able to predict the completion time of coexisting workloads within a 16% error for realistic workloads.

Verma *et al.* [36] performed an experimental study about application performance isolation, virtualization overhead with multiple VMs, and scenarios where applications were isolated from each other. From the insights obtained, authors proposed a framework and methodology for power-aware application placement for high-performance computing (HPC) applications, which considered both CPU and cache size constraints. Some of those important insights include (i) isolation on virtualized platforms works for homogeneous workloads (in terms of size) as long as they are very small, and the sum of them is inferior than cache size, or very large (e.g., 60 MB); (ii) performance tends to be better when the number of VMs is a power of 2, because memory banks as well as cache segments are typically power of 2; (iii) the performance of an application degrades as soon as it gets closer to the size of the cache resource, and stabilizes in some point after; and (iv) the power drawn increases with the applications working set, up to the full cache usage, and decreases after that because the CPU throughput decreases as well due to memory constraints. This last finding was confirmed later by Ref. [102]. However, authors work is based only on contention over CPU and LLC resources, and the approach does not consider autoscaling requirements of applications, nor readjust the provisioning of resources in the case of nonobservation of applications' QoS requirements.

These studies contributed to characterize the performance interference among co-hosted VMs that compete for shared resources, and are the basis for several state-of-the-art scheduling strategies to optimize energy efficiency in cloud data centers, while the QoS of running applications is met.

5. CURRENT RESOURCE MANAGEMENT IN CLOUD DATA CENTERS

As introduced in Section 2.3, energy consumption in a data center is mainly dictated by power consumed in servers. This section presents an overview about research works on techniques and strategies to manage computing resources in a energy-efficient way, and presents important remarks achieved.

5.1 Power- and SLA-Based Management of Resources

Despite the considerable efforts toward building green facilities equipped with advanced cooling systems [5, 21, 103] with a PUE near unity, it is evident that more attention has to be given to energy consumption by servers.

Because average resource utilization in most data centers can be as low as 10%, and an idle server can consume as much as 70% of the power consumed when it is fully utilized [35], idle servers represent a major concern (even AC to DC conversion losses in power supplies of computer systems implies more loss in cooling). A broad conclusion is that significant energy savings can be achieved in cloud computing through techniques such as VM consolidation, which explore the dynamic characteristic of workloads submitted to the cloud. The problem is seen as the traditional NP-hard bin-packing problem, in which several dimensions can be considered (e.g., CPU, RAM, network bandwidth, and disk). Greatly empowered by virtualization technologies that facilitate the running of several applications on a single physical resource concurrently, dynamic turning on and off servers method can be applied to reduce overall idle power draw, hence improving the energy efficiency. In this regarding, some research simply uses migration to aggregate the load on some servers and switch off idle free ones, in order to improve energy efficiency [104]. Despite the impressive consumption peak, due to start of all the fans and the ignition of all the node components, by booting physical servers that have been switched off, an on/off algorithm is still more energy efficient.

Several studies based on consolidation of VMs and dynamic switch on/off of PMs, by exploiting the variation in workload demands, have been proposed in order to improve energy efficiency. For example, Beloglazov et al. [61] developed allocation heuristics provision computing resources to client applications (e.g., web applications) with the aim of improving energy efficiency of the data center, while guaranteeing required QoS. The strategy implies to consolidate VMs in the fewest number of physical servers, leaving idle ones to go to sleep. Authors modeled the power consumed by a server based on linear power equation, as expressed by Eq. (5) in Section 2.5. The strategy to allocate VMs is based on the Best-Fit Decreasing heuristic (BFD), and it works by iteratively picking up the VM currently utilizing more CPU and allocating it in the physical server leading to the least increase of power consumption. In order to improve the current VM allocation, and thus to optimize the energy efficiency and guarantee the QoS expectations, authors propose three different strategies based on CPU utilization of a physical server. To this end, whenever the CPU utilization of a physical server falls outside the specified double-threshold interval, the strategy selects VMs that need to be migrated which then are placed, one by one, on the hosts using the modified BFD algorithm mentioned above. The results have shown that selecting the minimum number of

VMs needed to migrate from a host performs better in terms of QoS constraints and optimizes more energy efficiency compared to DVFS. Lee and Zomaya [63] conducted experiments over two proposed energy-conscious task consolidation heuristics. The main idea is to reduce the number of active resources so as to minimize the energy consumption for executing the task, without any performance degradation (i.e., without violating the deadline constraint). By dynamically consolidating tasks in fewer resources, idle physical servers can enter into power-saving mode or even be turned off. Proposed heuristics rely on the optimization of energy consumption derived from Eq. (5), meaning that energy consumption is directly proportional to resource utilization. Another alternative is vGreen [62], a framework to manage VMs scheduling across different physical servers with the objective of managing and improving the overall performance and system-level energy savings. vGreen exploits internal characteristics of VMs (i.e., in terms of instructions per cycle, memory accesses) and their performance and power consumption. The approach presents a client–server model: each physical server runs a client instance, which performs online characterization of the executing VMs and sends update information to a central server. In turn, the central server has the duty of scheduling VMs across physical servers. The scheduling algorithm works by co-locating VMs with heterogeneous characteristics, by balancing three specific metrics that account for memory accesses per clock cycles, instructions per clock cycle, and CPU utilization for every VM. In the end, idle nodes in the system can be moved into low power states.

Some approaches have considered the optimization of several objectives at the same time. In fact, power consumed by physical servers, thermal dissipation, and maximization of resource usage are major concerns that have been subject of significant study. In this regarding, Xu and Fortes [30] proposed a method to optimize conflicting objectives associated to initial virtual resource management. The size of a VM is represented as a d–dimensional vector in which each dimension corresponds to one type of the requested resources (e.g., CPU, memory, and storage). Considered objectives are reduction of resource waste, operational power, and mitigation of individual hotspots by keeping temperature within a safe operating range. The problem of VM placement is formulated as a multi-objective combinatorial optimization problem that is solved as two-level control approach: (i) at first level, controllers implemented in every VM are responsible for determining the amount of resources needed by an application and (ii) in the second level, a global controller determines VM placement to physical resources.

A grouping genetic algorithm [105] is utilized for combinatorial optimization (i.e., VMs to PMs mapping), and the solutions obtained are then evaluated using a fuzzy logic system to optimize the several objectives. Later, same authors [106] complemented the first work with a dynamic mapping of VMs to physical resources solution, which reassigns VMs to hosts due to changes of system conditions or VMs requirements. The objectives to achieve include elimination of thermal hotspots, minimization of total power consumption, and achieving desired application performance. First, the system analyzes samples related to objectives within a temporal window for each PM, and if sample values are beyond predefined thresholds, then the system reconfigures VMs to PMs mapping by maximizing an utility function that combines all normalized objectives. The proposal was evaluated experimentally using a mix of types of workloads to emulate the variety and dynamics of data center workloads. Authors have shown that their approach significantly reduces unnecessary VM migrations by up to 80%, avoids unstable host selection, and improves the application performance by up to 30% and the efficiencies of power usage by up to 20%.

Several of the above solutions do not consider all the resources utilized by a VM (i.e., CPU, RAM, network bandwidth, and disk). In order to tackle this limitation, Mastroianni et al. [64] proposed the ecoCloud project, a self-organizing and adaptive approach for the consolidation of VMs on resources. Unlike most of the previous approaches, ecoCloud considers multiple-resource dimensions (e.g., CPU and RAM) when taking decisions on the assignment and migration of VMs to physical resources. The workload of each application modeled as a VM is dynamic, implying for example that CPU demand varies with time. The aim is to increase the level of utilization of servers by applying dynamic consolidation of workloads, with the twofold objective of increasing energy efficiency and maintaining the SLA stipulated with users. The decisions about configuration of VMs to physical servers mapping are, in a first stage, taken locally on each physical server where it is decided whether a new VM is accepted or rejected according to the differential between available and demanded resources, and then properly combined by the data center manager, at a high decision level. Because the problem is so complex to solve, and hardly scalable, assignment and migration of VMs are driven by ant algorithms [107] which consider local information exclusively.

Other alternatives explore consolidation of workloads with dissimilar requirements. In this sense, Zhu et al. [65] developed pSciMapper, a power-aware consolidation framework for scientific workflows in heterogeneous cloud resources. It exploits the fact that tasks of a given workflow can have substantially different resource requirements and can vary along time. First, pSciMapper analyzes the correlation between workloads with different resource usage profiles (i.e., CPU, RAM, disk, and network bandwidth), and how power and performance are impacted by their co-location. Then, it consolidates workloads with dissimilar requirements as a measure to reduce total power consumption and resource requirements simultaneously. Authors start by cluster/group dissimilar workflows based on the interference between the resources they require. Then, clusters obtained are mapped to the given set of servers by using the Nelder–Mead [108] algorithm. During this mapping phase, the kernel canonical correlation analysis method is used to relate resource requirements to performance and power consumption. The algorithm developed can greatly reduce power consumption with lower execution slowdown. An important outcome from experiments conducted relates to the fact that using nonwork-conserving mode by setting the cap parameter in Xen credit scheduler saves power (this outcome is leveraged later by Refs. [109, 110]).

In an effort to improve resource utilization and optimize energy efficiency in cloud realistic scenarios which comprise different types of workloads with dissimilar QoS constraints, Garg et al. [60] proposed a dynamic resource management strategy to handle scheduling of two types of applications, namely, compute-intensive noninteractive jobs and transactional applications such as web server. The mechanism responds to changes in transactional workloads and tries to maximize resource utilization while enforcing the SLA requirements and specifications. A web VM is scheduled based on the best-fit manner, not taking into account the resources consumed by VMs running dynamic jobs. In order to maximize the energy efficiency, the scheduler tries to explore free resources in nodes running web applications to schedule dynamic jobs. However, if free resources are not sufficient so the task can be executed within its deadline, then the job is allocated in another node in the form of static job. Resources are transferred from dynamic jobs to web applications as demand changes. Periodically, the algorithm verifies the SLA accomplishment and realizes consolidation. The priority of scheduling is (1) enforce SLAs, (2) schedule the jobs from batch job queue, and (3) consolidation. The limitation of this work is that authors do not consider the case in which consolidation of workloads may exceed the total available resources in the host, during runtime.

Unlike previous approaches, which realize online/dynamic scheduling of VMs to improve energy efficiency via consolidation, Verma et al. [111]

analyzed the associated risks due to static consolidation of applications. The approach studies workloads characteristics and utilizes correlation between workloads for determining the most effective static consolidation configuration. Using the insights from the workload analysis, authors proposed two consolidation strategies, namely Correlation-Based Placement and Peak Clustering-based Placement. Basically, the techniques exploit the correlation between applications in terms of resource needs along time, and peak demand. Authors show that the strategies perform better in terms of power savings and SLA infringements according to the type of servers, application size, and number of applications. Despite static approaches impose less overhead than dynamic/online approaches (e.g., due to migration and consolidation of VMs in less number of PMs to improve energy efficiency), they imply previous knowledge of application workload characteristics. As such, the approach can reveal hard to apply to cloud computing environments due to the wide spectrum of workload characteristics.

5.2 Power- and Failure-Aware Management of Resources

As mentioned in Section 4.2, failure events also contribute to increase the energy waste in large-scale data centers. In a perspective to tackle this issue, Fu and Xu have investigated for some years in the field of PFT and have contributed with relevant methodologies in the correlation and prediction of failures [82, 85]. Based on these contributions, the authors have later proposed a failure- and power-aware resource management approach, by means of VM migrations, in order to enhance systems availability and to reduce energy consumption [112]. Basically, the authors propose two algorithm alternatives in the scheduling of deadline-driven jobs (a job is defined as a set of tasks) that consider both the capacity and reliability status of compute nodes in the process of making selection decisions in provisioning of resources. The first scheduling alternative, Optimistic Best-Fit (OBFIT) algorithm, schedules jobs in nodes that are not predicted to fail before the jobs' deadline and cumulatively provide the minimum required resources to execute the jobs within their deadlines. The second alternative, called Pessimistic Best-Fit algorithm, calculates the average available capacity level $C_{average}$ among the compute nodes that will not fail before job deadline, and from a set of PMs that will fail before user job deadline, it selects the PM with capacity C_p such that $C_{average} + C_p$ results in the minimum available capacity to run a VM. Since both algorithms are best-fit type, the power consumption is optimized, and the waste of energy due to failures is reduced because the algorithms are failure-aware. From experiences, authors show that system productivity can be enhanced by considering the reliability of nodes, as the job completion rate is increased by 17.6%, and the task completion rate reaches 91.7% with 83.6% utilization of relatively unreliable nodes. The limitation of Fu approach is to assess the risk of failure only at the time of (re)scheduling, and to consider that a VM will suffer a single failure during execution.

Some authors leveraged PFT in the scheduling of workflows. According to Arabnejad and Barbosa [113], a workflow can be defined as a set of interconnected steps that must be executed to accomplish some work. Usually, a workflow is described by a Directed Acyclic Graph, where the vertices are the tasks and the edges are the temporal relations between the tasks. Because workflows constitute a common model for describing a wide range of applications in distributed systems, several projects have designed workflow management systems to manage the execution of applications in distributed systems. In this regarding, Yusuf et al. [114] started by proposing recovery-aware service components (i.e., self-contained autonomous processes such as server processes running in a VM), which combine prediction with reactive and PFT techniques to improve the reliability of workflow applications. It works by preemptively rebooting components to probabilistically safer states whenever probability of failure goes beyond a threshold. Later in Ref. [115], the same authors extended the work to support MapReduce workflow applications. Both works targeted the grid. In Ref. [116], Yusuf and Schmidt have analyzed their last contribution in a cloud computing environment and also evaluated the case of combinational logic architectural patterns which includes, for example, massive real-time streaming, big data processing, or scientific workflows. The results allowed authors to conclude that their fault tolerance approach improves systems reliability compared to using only prediction-based fault tolerance or reactive fault tolerance.

The literature is extensive in contributions in the field of fault tolerance (e.g., Refs. [117–122]). However, these approaches consider fault tolerance at all cost, discarding the objective of energy efficiency.

5.3 Interference-Aware Management of Resources

The extensive research in understanding and modeling performance interference effects among consolidated workloads has shed significant light on the kind of settings and combinations that most degrade applications performance. Based on the outcomes from these studies, several authors have proposed alternative QoS- or performance-aware scheduling algorithms in order to achieve distinct objectives, and guaranteeing applications required QoS at the same time.

In this regarding, Kim *et al.* [123] proposed a Performance-Maximizing VM scheduler algorithm that deals with performance impact due to contention in a shared LLC. Authors showed that the performance degradation can reach almost 43% due to interference in shared LLC and memory bus. The aim was to identify the VM arrangement that minimizes the performance interferences and the number of active physical servers, thus being possible to reduce energy consumption by turning off redundant servers. The proposal implies to co-locate a VM that has large cache demand with a VM that has small cache demand, in order to improve overall performance. Cache demand for a VM is inferred from LLC reference ratio and can be determined when VM is running, without the need of prior profiling of workloads. The scheduling algorithm follows this strategy, allocating VMs with the largest LLC reference ratio in PMs in which the sum of LLC reference ratios is the smallest. Results have demonstrated a decrease in average performance degradation ratio of about 16%, even though average performance degradation ratio remains above 14%. Chiang and Huang [124] proposed Tracon framework so as to mitigate the interference effects from co-located data-intensive applications, and thus improving the overall system performance. Authors focus on a specific I/O type of resources. Tracon is composed of three major components, namely: (i) interference prediction component to infer application's performance from resources consumption; (ii) an interference-aware scheduler to utilize the first component and assign tasks to physical resources in an optimized way; and (iii) and task and resource monitors to collect application characteristics to feed both previous components. The framework works based on modeling and control techniques from statistical machine learning, for reasoning about the application's performance under interference to manage the virtual environment. In this way, the application performance is inferred from an interference prediction model which is based on the resource usage. Authors test three modeling approaches and three scheduling algorithms. Results showed that Tracon can improve application runtime and I/O throughput up to 50% and 80%, respectively. Nathuji *et al.* [125] proposed Q-Clouds, which utilizes online feedback to build a multi-input multi-output model to capture performance interferences among consolidated CPU-bound workloads. Then, it reacts to performance degradation by adjusting processor allocation for each application based on the required SLA. Through the hypervisor, it sets the cap mechanism on each VM to control the processing resources an application can use. It works by maintaining free resources to be lately used in the adjustment, without the need to determine the underlying sources of interference. Furthermore, the authors use Q-States to differentiate various levels of SLA so that they can dynamically provision underutilized resources, thereby improving system efficiency. Results show that performance interference is mitigated completely when resources are available, and the use of Q-States improves system utilization by up to 35%. Q-States vary from minimal performance that an application requires, up to high levels of QoS defined by the customers willing to pay for them. Zhu and Tung [91] proposed an interference model which predicts the impact of application performance due to interference from different levels of shared resources (e.g., on-chip, storage, and network) under contention. Authors proposal consider time variance in application workloads and use an influence matrix to estimate the additional resources required by an application to achieve desired QoS, defined in terms of execution time, as if it was executed alone. Then, leveraged by the proposed model to realize better workload placement decisions, authors introduce a consolidation algorithm to optimize the number of successfully executed applications in the cloud. The average prediction error is stated as being less than 8%. As a result, the number of successfully executed applications is optimized, outperforming static optimal approaches and a state-of-art approach such as Q-Clouds [125], with negligible overhead. The approach implies a previous training period to profile the applications' behavior when running co-located with others.

In a recent study, Sampaio and Barbosa [126] developed PIASA, a power- and interference-aware strategy to create and manage IaaS in the cloud. The aim is to manage resource allocation within a data center that runs CPU- and network I/O-bound applications. While CPU-bound applications can suffer performance degradation due to contention in on-chip resources, bursty variation in resource demands of network I/O-bound applications leads to deviations from required performance. Authors proposed two mechanisms, based on Kalman filters, that are able to estimate the performance deviation in both types of workloads, and developed a scheduling algorithm that compensates deviations from required performance in both types of applications. The scheduling algorithm works by applying readjustments in the VMs to PMs mapping to correct such performance deviations. At the same time, the refinement in the assignment of resources to VMs considers the optimization of energy efficiency of the underlying infrastructure. PIASA allows cloud providers to construct IaaS to end-users that are energy-efficient-bound, performance-bound, or balanced in terms of energy efficiency and performance, according to their needs and objectives.

6. A PRACTICAL CASE OF ENERGY-EFFICIENT AND SLA-BASED MANAGEMENT OF RESOURCES

This section presents a study about a multi-objective approach to energy-efficient and high-available dynamic and reconfigurable distributed VM infrastructure (i.e., IaaS) for cloud computing systems. It follows a previous study about energy-efficient and fault tolerance IaaS, properly described in Ref. [110]. In this approach, the cloud resource manager efficiently utilizes system resources for high-availability computing (through PFT technique) and energy efficiency optimization to reduce operational costs and carbon footprints to the environment. The resource manager combines a power-optimizing mechanism that detects and mitigates power inefficiencies [127], a failure prediction tool [112] to foresee the next failure occurring in a physical server, and executes a power- and failure-aware decision making algorithm that leverages virtualization features (i.e., VM migration and establishment of a CPU resource limit assigned to each VM), to dynamically construct and readjust energy-efficient virtual clusters that execute users' jobs with a high level of SLA accomplishment.

6.1 Description of the Cloud Computing Environment

The target of this study is the management of an IaaS, where the cloud provider infrastructure is composed of a set of homogeneous physical servers in terms of hardware and software. The cloud middleware comprises a cloud manager, which receives monitoring data about resource usage, power efficiency, and reliability levels of the computing nodes. The cloud manager architecture is depicted in Fig. 4. End-users submit sets of jobs, in which each job comprises a set of independent deadline-driven tasks, intensive in terms of the CPU resource. The characteristics of submitted jobs and failure events are described in Sections 4.1 and 4.2, respectively. Based on monitoring data and end-users requests, the cloud manager dynamically constructs and manages energy-efficient and high-available virtual clusters to execute the sets of independent tasks in which computing resources are affected by failures. The cloud provider is application-agnostic, i.e., unaware of applications and workloads characteristics served by the VMs

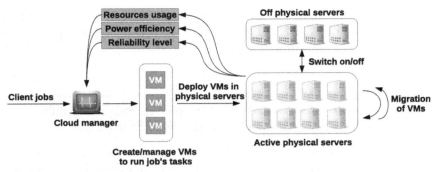

Figure 4 Cloud management architecture.

of the virtual clusters. The SLA, or contract, is defined for each job by specifying a deadline guaranteed by the system, which equals the deadline of its longest task. This study measures the service performance based on the success rate of job completion by the deadline.

6.2 Establishment of Metrics to Assess the Scheduling Algorithms

In order to evaluate and compare the performance of the scheduling algorithms, three metrics are defined. The first one, the completion rate of users' jobs E_J, is expressed by Eq. (11). It measures the completion rate of jobs, which is determined as the ratio of the number of jobs completed by their deadline, J_C, to the number of submitted jobs, J_S. A job is considered completed if all its tasks are completed by their deadlines. The value of metric E_J is comprehended within the interval $[0, 1]$, and the difference between E_J and 1, multiplied by 100, gives the percentage of SLA violations.

$$E_J = \frac{J_C}{J_S} \tag{11}$$

The second metric is the energy efficiency, E_M, as shown in Eq. (13). It measures the amount of energy consumed, in Joules, to produce useful work, which in turn is calculated as the number of Mflops associated with successfully completed jobs only, J_C. In turn, the energy consumption is determined by multiplying the average power consumption of the computing infrastructure (i.e., for all active physical nodes u at all sample times f) by the number of sample times f, multiplied by 60 (since power consumption samples are obtained each minute). In the end, E_M is calculated by dividing the sum of the workloads from all tasks of successfully completed jobs by the overall energy consumption. Hence, the energy efficiency metric is represented as Mflops/J.

$$E_M = \frac{\sum_j (\theta_j \times \sum_{t=1}^{n} W(t,0))}{\frac{\sum_{s=1}^{f} \frac{\sum_{i=1}^{u} P_i}{u}}{f} \times f \times 60}, \quad \theta_j = \begin{cases} 1, & \text{if job } j \text{ completed} \\ 0, & \text{otherwise} \end{cases} \quad (12)$$

The third metric E_W, introduced in Eq. (13), represents the working efficiency, E_W. This metric is used to determine the quantity of useful work performed (i.e., the completion rate of users' jobs) by the consumed power. It is calculated by multiplying E_J in Eq. (11) by the average power efficiency E_{P_i} defined by Eq. (7), for all active physical nodes $i \in [1,u]$ at all sample times f. By efficiently consolidating the VMs, the average power efficiency of the execution of all jobs can be optimized, for all active physical nodes u.

$$E_W = \frac{\sum_{s=1}^{f} \frac{\sum_{i=1}^{u} E_{P_i}}{u}}{f} \times E_J, \quad \forall u \le h \quad (13)$$

Equations (12) and (13) capture the amount of useful work performed at two different levels. While the first metric quantifies the number of useful Mflops by the consumed energy, the second measures the quantity of useful work, in terms of jobs successfully completed by the deadline, performed with the consumed power. The best algorithm should optimize the processing of more Mflops with a lower amount of energy and maximize the job completion rate while keeping high levels of power efficiency.

6.3 Strategies Applied to Improve Energy Efficiency

Three scheduling algorithms to create dynamic maps of VMs to PMs are tested in this study. The first algorithm, Common Best-Fit and here henceforth called CBFIT, is a best-fit-type algorithm that is based on the MBFD algorithm presented by Beloglazov et al. [61]. The CBFIT strategy selects, from all available PMs, the PM that has the minimum capacity necessary to run a task to optimize energy consumption. Because MBFD does not consider the reliability of the nodes, CBFIT is a simplified version to show the impact of the proactive fault-tolerant migrations. The OBFIT algorithm is a failure- and power-aware scheduling algorithm proposed by Fu [112] and properly discussed in Section 5.2. POwer- and Failure-Aware Relaxed time Execution (POFARE) is a power- and failure-aware scheduling algorithm proposed by Sampaio and Barbosa [110]. Basically, the algorithm starts by creating a list of tasks to schedule, from the highest to the lowest priority: (1) proactive failure tolerance, (2) reinitiating of failed tasks, and (3) scheduling of new tasks. Power-optimizing scheduling has the least priority, being performed when there are no other tasks. Then, the destination PM is selected from all the PMs in the infrastructure as the one that cumulatively (i) provides enough CPU capacity to run the task within the deadline; (ii) optimizes more the power efficiency (properly defined by Eq. (7) in Section 2.5); and (iii) presents itself as the most reliable (i.e., biggest MTBF). In the end, the algorithm sets the cap parameter in the Xen credit scheduler for fine-grained CPU assignment, allocating the strictly necessary amount of resources to complete the tasks within their deadlines. The objective is to increase the level of consolidation of VMs and consequently reduce the number of active PMs. In every strategy, the energy efficiency is improved through dynamic VM consolidation, which, based on a power efficiency detection mechanism [127], transfers VMs from lower loaded PMs to other PMs, and by putting the first PMs in sleep mode. Replacement of VMs is achieved through stop and copy migration technique, which according to Fu [112], and Sampaio and Barbosa [110], is simpler and faster to accomplish, despite introducing superior service downtime. Such properties are more preferable in failure-aware resource management schemes due to small overhead introduced and the imperfection of state-of-the-art failure prediction techniques.

6.4 Simulation Results and Analysis

The simulated cloud computing infrastructure comprised 50 homogeneous physical nodes, each of which having the CPU capacity of 800 Mflops/s. The power consumed by a PM is modeled by Eq. (5), in which $P_{max} = 230$ W and $P_{idle} = 0.7 \times P_{max}$. These values are in line with what is specified by the SPECpower benchmark [128] for the first quarter of 2014, for modern servers. The network bandwidth was set to 1 GB/s. Regarding memory usage, a VM requires a RAM size of 256, 512, or 1024 MB, randomly selected. The single migration of a VM takes 8, 10, or 12 s to accomplish, which is in line with Ref. [112], and it depends on the memory size and the free network bandwidth. A task runs in a single VM at a time. Consolidation of VMs is performed when a PM is detected with a CPU usage below 55% in three of five CPU usage samples, which are taken in intervals of 1 min. The failure prediction tool is able to foresee when a PM is going to fail with 75% of average accuracy. The failure prediction accuracy (FPA) of the failure prediction tool is defined as the difference between the actual and predicted failure time, expressed as a percentage, and is assumed that a failure is predicted before occurring. VMs are migrated away from a PM that is predicted to fail 3 min before the predicted time of failure.

Based on the workload characteristics introduced in Section 4.1, which derived from a extensive analysis to the Google Cloud tracelogs, a set of 3614 synthetic jobs was created to simulate cloud users' jobs. Each task requires a RAM size of 256, 512, or 1024 MB, selected randomly. The jobs comprised a total of 10,357 tasks. Each task deadline was rounded up to 10% more than its minimum necessary execution time. Similarly, failure characteristics for PMs followed the characteristics indicated in Section 4.2, with a MTBF of 200 min, and MTTR with a mean time set to 20 min, varying up to 150 min.

The aim of the first set of simulations is to assess the advantages in utilizing dynamic scheduling of VMs, which applies consolidation of VMs to improve the energy efficiency of the cloud infrastructure. Physical servers are not subject to failure. Table 3 registers the results of the simulation, with (wcs) and without (wocs) consolidation of VMs. The numbers for the energy consumption (i.e., "Energy (MJ)" column) demonstrate a reduction of 9.4% for CBFIT, 9.0% for OBFIT, and 11.2% for POFARE when consolidation of VMs is applied. In this regarding, the average number of VMs per PM (i.e., "Av. # of VMs/PM" column) increases as a result of consolidation. Moreover, consolidation of VMs is achieved through migration, which explains the increment in the number of migrations of VMs (i.e., "VM Migrations (%)" column) with consolidation. Comprehensibly, one can observe that improving the energy efficiency negatively impacts the rate of completed jobs (i.e., "E_J (%)" column), and consequently the rate of SLA violations. However, such impact is residual of less than 0.1% for OBFIT and POFARE scheduling algorithms. Comparing directly the algorithms, the numbers confirm POFARE as the best algorithm, consuming less energy for the same rate of completed jobs.

The next set of simulations intends to evaluate the use of consolidation of VMs and fault tolerance to improve the energy efficiency of cloud data centers. Table 4 shows the results for a FPA fixed in 75%, with and without consolidation mechanism. The results confirm POFARE as the best algorithm, achieving superior savings in energy consumption (i.e., "Energy (MJ)" column) and with a small impact of about 0.2% in the completion rate of jobs (with and without consolidation). With consolidation, POFARE

Table 3 Results for Google Cloud Tracelogs-Based Workloads Without Failures for Energy Consumption, Job Completion Rate, Ratio of VM Migrations to Total Number of Tasks, and Number of VMs per PM, Without Consolidation (wocs) and with Consolidation (wcs)

Algorithm	Energy (MJ) wocs/wcs	E_j (%) wocs/wcs	VM Migrations (%) wocs/wcs	Av. # of VMs/PM wocs/wcs
CBFIT	27.8/25.2	100/99.8	0/2.10	16/17
OBFIT	27.7/25.2	100/99.9	0/1.99	16/17
POFARE	26.9/23.9	100/99.9	0/1.82	21/22

Table 4 Results for Google Cloud Tracelogs-Based Workloads for Energy Consumption, Job Completion Rate, Ratio of VM Migrations to Total Number of Tasks, and Number of VMs per PM for the Case of 75% of Failure Prediction Accuracy Without Consolidation (wocs) and with Consolidation (wcs)

Algorithm	Energy (MJ) wocs/wcs	E_j (%) wocs/wcs	VM Migrations (%) wocs/wcs	Av. # of VMs/PM wocs/wcs
CBFIT	27.7/25.1	95/95.5	(3.12) + 1.6	16/17
OBFIT	28.3/27.2	100/100	0 / 0.99	17/16
POFARE	26.9/23.9	99.8/99.8	0.8/2.55	21/22

reduces the energy consumption by 11.2%, from 26.9 to 23.9 MJ, and keeping the same job completion rate. Moreover, consolidation of VMs implies increasing the number of migrations from 0.8%, due to failure tolerance migrations, to 2.55%, which also includes power-oriented migrations. In turn, CBFIT algorithm, which is not aware of PMs unavailability, degrades in performance with failure events, achieving the worst completion rate of jobs compared to other failure-aware scheduling algorithms OBFIT and POFARE. At the same time, the results show that with the consolidation mechanism, the energy consumption is reduced while the rate of finished jobs is improved, for all the algorithms.

Figure 5 presents the results for a set of simulations aiming at analyzing the influence of the FPA in the performance of the scheduling algorithms in improving the energy efficiency. In such scenario, VMs are consolidated, PMs are subject to failure, and FPA varies from 10% to 100%. Figure 5A reveals that CBFIT clearly suffers performance degradation in terms of rate of completed jobs, which is constant despite variation in FPA, since it does not consider failures. On the other hand, POFARE and OBFIT, which are

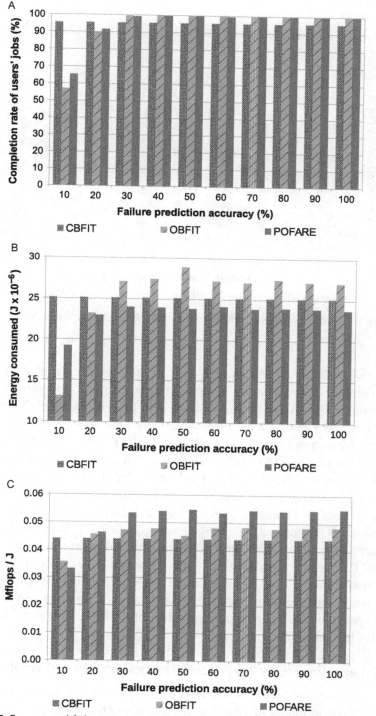

Figure 5 Energy- and failure-aware scheduling algorithms, with dynamic consolidation, for Google Cloud tracelogs-based workloads. (A) Completion rate of jobs regarding failure prediction accuracy. (B) Average energy consumed regarding failure prediction accuracy. (C) Energy efficiency regarding failure prediction accuracy.

(Continued)

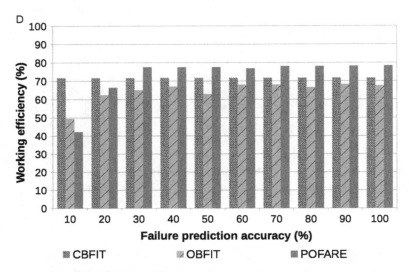

Figure 5—Cont'd (D) Working efficiency regarding failure prediction accuracy.

power- and failure-aware algorithms, are able to conclude more jobs, despite the failures occurring in the computing infrastructure. However, Fig. 5B shows that POFARE consumes about 9% less energy than OBFIT, in average, and 12% less for a FPA = 75% (i.e., the average FPA provided by the failure prediction tool [112]), for similar rate of completed jobs. In the specific case of FPA = 10%, OBFIT is unable to select PMs with enough reliability to run the tasks (i.e., recall that FPA affects the perception of MTBF), and consequently completes less number of jobs and consumes less energy than CBFIT and POFARE. Figure 5C and D presents the amount of useful work performed, in terms of energy efficiency and working efficiency. In both graphics, POFARE algorithm outperforms CBFIT and OBFIT strategies, processing more Mflops with a lower amount of energy and maximizing the job completion rate while keeping high levels of power efficiency. An interesting finding is that POFARE provides performance and system availability nearly independent of the FPA, only degrading for a FPA less than 30%.

6.5 Experiments in Real Cloud Testbed and Result Analysis

To confirm the results obtained above through simulation, this section evaluates the performance of the tested alternatives in a real platform. The experimental testbed consisted of 24 nodes (cores) from an IBM cluster with the following characteristics: each physical server was equipped with an Intel Xeon CPU E5504 composed of 8 cores working at 2.00 GHz, supporting virtualization extensions and deploying 24 GB of RAM. The Xen 4.1 virtualization software was installed in the Ubuntu Server 12.04 operating system. The cloud manager is aware of nodes' status by collecting heartbeat messages [129]. A failure is considered when a node stops sending heartbeat messages. The real platform configuration follows the same configuration of the simulation scenario. The experiments took place for a FPA varying within {25,50,75}%. The characteristics of the jobs and failures follow those used in simulations. A set of 151 jobs with a total of 389 tasks was created to run for approximately 1 h to complete an instance of the experiment.

Figure 6 shows the results obtained from the experiments when applying both energy-optimizing and fault-tolerant mechanisms to drive decisions on the assignment and migration of VMs. These results confirm, in general, that POFARE is the best algorithm to optimize the energy efficiency and to complete a similar number of jobs by their deadlines for FPA values equal to or greater than 25%, as pointed out by simulation results (Fig. 5C and D). In this sense, power- and failure-aware strategies are better to improve energy efficiency than other alternatives that only perform consolidation of VMs. In the particular case of FPA = 100% in Fig. 6A, the delivered Mflops/J by the POFARE algorithm decreases due to the fluctuation of the assigned amount of CPU to each task, which is not constant in the real system, as considered in the simulation. Such effect no longer affects the performance of POFARE in the case of lower values of FPA because the error margin used to avoid failures accommodates such fluctuations.

6.6 Conclusions

To maximize profit, cloud providers need to apply energy-efficient resource management strategies and implement failure tolerance techniques. Dealing with failures not only increases the availability of a system and services perceived by cloud users but also has serious implications in energy waste. However, reducing energy consumption through consolidation and providing availability against failures are not trivial, as they may result in violations of the agreed SLAs.

This section demonstrated how decisions on the assignment and migration of VMs, driven by power efficiency and PFT, can improve the energy efficiency of a cloud computing infrastructure. In this regarding, three different scheduling algorithms were tested and their performance analyzed.

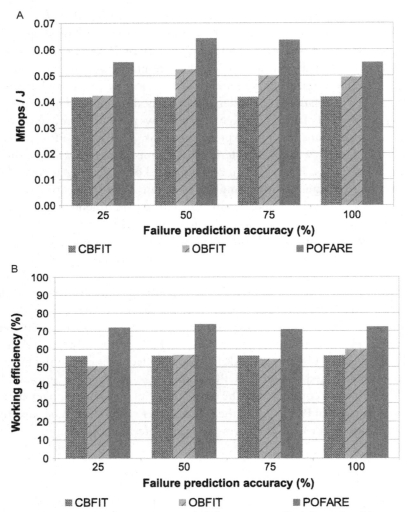

Figure 6 Experiments for energy- and failure-aware scheduling algorithms, with dynamic consolidation, for Google-based workloads. (A) Energy efficiency regarding failure prediction accuracy. (B) Working efficiency regarding failure prediction accuracy.

The results have shown that power- and failure-aware scheduling algorithms are the best alternatives to cope with node failures and consolidate VMs to optimize energy efficiency. In particular, scheduling algorithms which assign the strict amount of required CPU resources so tasks can finish by their deadlines, as is the case of POFARE, are able to achieve higher levels of consolidations, thus simultaneously increasing more the amount of useful Mflops processed per energy unit, as well as the number of jobs completed by the consumed power.

7. CONCLUSIONS AND OPEN CHALLENGES

Cloud computing providers are under great pressure to reduce operational costs and increase the return on investment. Energy consumption in data centers represents a very important and significant fraction of the whole cost of operation. The problem is even more exacerbated as cloud data centers infrastructures continue to grow in size and in complexity, to respond to consumer demands, because failure events become norm instead of exception, increasing even more the energy waste. Tightly coupled with energy consumption, data centers produce significant CO_2 emissions into the environment and contributes for the global warming.

In order to reduce the power and energy consumption in data centers, it is of paramount importance to adopt new strategies that efficiently manage the computing resources, in a power- and failure-aware manner. This chapter has contributed with an overview about existing techniques and strategies for optimization of power and energy in cloud data centers, achieved via consolidation of applications and PFT in physical nodes.

Due to the complexity of data center infrastructures in deploying and configuring resources in a power- and energy-efficient way, and fulfilling with diverse constraints at same time, several key problems remain open. For example, despite consolidation of VM instances on less number of physical servers contributes to improve the utilization of resources and to reduce costs related to electricity bills, dynamic models able to predict the energetic behavior and the economic analysis of data centers are needed. Even though VM consolidation contributes to improve energy efficiency in the data center, limitations in virtualization technology can conduct to interference among VM instances in a physical server, which may degrade the performance of applications running on those VMs.

Regarding data center facilities, one need to develop efficient models capable of dynamically estimate the energy consumption and mass flows into the data centers to improve their characterization to facilitate further developments in the area. Complementary measures that consider a holistic view of power consumption by communication, storage, cooling, and servers components are needed and can be handled via multi-objective approaches.

Much of the results are preliminary, in the sense that many of the current strategies are evaluated via simulation. Therefore, results from tests in real-case scenarios are needed to better assess the performance and eventually mitigate some limitations of the models and algorithms proposed.

REFERENCES

[1] A.S. Tanenbaum, M. Van Steen, Distributed Systems: Principles and Paradigms, second ed., Prentice-Hall, Inc., Upper Saddle River, NJ, 2006, ISBN 0132392275.
[2] T. Erl, Service-Oriented Architecture: Concepts, Technology, and Design, Prentice Hall PTR, Upper Saddle River, NJ, 2005, 0131858580.
[3] G. Vallee, T. Naughton, C. Engelmann, H. Ong, S.L. Scott, System-level virtualization for high performance computing, in: 16th Euromicro Conference on Parallel, Distributed and Network-Based Processing, 2008, PDP 2008, IEEE, 2008, pp. 636–643.
[4] M. Armbrust, A. Fox, R. Griffith, A.D. Joseph, R. Katz, A. Konwinski, G. Lee, D. Patterson, A. Rabkin, I. Stoica, et al., A view of cloud computing, Commun. ACM 53 (4) (2010) 50–58.
[5] Purdue University, RCAC, Carter cluster. https://www.rcac.purdue.edu/userinfo/resources/carter/, 2011, (accessed 25.03.14).
[6] N.R.C. (US), The Rise of Games and High-Performance Computing for Modeling and Simulation, National Academies Press, Washington, DC, 2010.
[7] G.E. Moore, et al., Cramming more components onto integrated circuits, Proc. IEEE 86 (1) (1998) 82–85.
[8] J.G. Koomey, Estimating total power consumption by servers in the US and the world, 2007.
[9] J. Koomey, Growth in data center electricity use 2005 to 2010, A report by Analytical Press, completed at the request of The New York Times, 2011, p. 9.
[10] J.M. Kaplan, W. Forrest, N. Kindler, Revolutionizing Data Center Energy Efficiency, Technical Report, McKinsey & Company.
[11] E. Commission, et al., Code of conduct on data centres energy efficiency: version 1.0, 2010.
[12] G. Von Laszewski, L. Wang, A.J. Younge, X. He, Power-aware scheduling of virtual machines in DVFS-enabled clusters, in: IEEE International Conference on Cluster Computing and Workshops, 2009, CLUSTER'09, IEEE, 2009, pp. 1–10.
[13] V. Venkatachalam, M. Franz, Power reduction techniques for microprocessor systems, ACM Comput. Surv. 37 (3) (2005) 195–237.
[14] R. Brown, et al., Report to Congress on Server and Data Center Energy Efficiency: Public Law 109-431, Lawrence Berkeley National Laboratory, 2008.
[15] O. VanGeet, FEMP Best Practices Guide for Energy-Efficient Data Center Design, National Renewable Energy Laboratory, 2011.
[16] A.Y. Zomaya, Y.C. Lee, Energy Efficient Distributed Computing Systems, vol. 88, John Wiley & Sons, Hoboken, NJ, 2012.
[17] C. Leangsuksun, T. Rao, A. Tikotekar, S.L. Scott, R. Libby, J.S. Vetter, Y.-C. Fang, H. Ong, IPMI-based efficient notification framework for large scale cluster computing, in: Sixth IEEE International Symposium on Cluster Computing and the Grid, 2006, CCGRID 06, vol. 2, IEEE, 2006, p. 23.
[18] M.D. De Assuncao, J.-P. Gelas, L. Lefevre, A.-C. Orgerie, The Green Grid'5000: instrumenting and using a grid with energy sensors, in: Remote Instrumentation for eScience and Related Aspects, Springer, New York, 2012, pp. 25–42.
[19] B. Lannoo, et al., Overview of ICT energy consumption. Deliverable D8.1, 2013.

[20] J. Cho, B.S. Kim, Evaluation of air management system's thermal performance for superior cooling efficiency in high-density data centers, Energy Build. 43 (9) (2011) 2145–2155.

[21] J. Kim, M. Ruggiero, D. Atienza, Free cooling-aware dynamic power management for green datacenters, in: 2012 International Conference on High Performance Computing and Simulation (HPCS), IEEE, 2012, pp. 140–146.

[22] E. Pakbaznia, M. Ghasemazar, M. Pedram, Temperature-aware dynamic resource provisioning in a power-optimized datacenter, in: Proceedings of the Conference on Design, Automation and Test in Europe, European Design and Automation Association, 2010, pp. 124–129.

[23] Apple, Environmental responsibility. http://www.apple.com/environment/climate-change/, 2015 (accessed 09.02.15).

[24] L.A. Barroso, The price of performance, Queue 3 (7) (2005) 48–53.

[25] L.A. Barroso, U. Hölzle, The datacenter as a computer: an introduction to the design of warehouse-scale machines, Synth. Lect. Comput. Archit. 4 (1) (2009) 1–108.

[26] A.-C. Orgerie, M.D. de Assuncao, L. Lefevre, A survey on techniques for improving the energy efficiency of large-scale distributed systems, ACM Comput. Surv. 46 (4) (2014) 47.

[27] M. Mazzucco, D. Dyachuk, R. Deters, Maximizing cloud providers' revenues via energy aware allocation policies, in: 2010 IEEE 3rd International Conference on Cloud Computing (CLOUD), IEEE, 2010, pp. 131–138.

[28] S. Rivoire, P. Ranganathan, C. Kozyrakis, A comparison of high-level full-system power models, HotPower 8 (2008) 3.

[29] X. Fan, W.-D. Weber, L.A. Barroso, Power provisioning for a warehouse-sized computer, ACM SIGARCH Comput. Archit. News 35 (2007) 13–23.

[30] J. Xu, J.A. Fortes, Multi-objective virtual machine placement in virtualized data center environments, in: Green Computing and Communications (GreenCom), 2010 IEEE/ACM Int'l Conference on & Int'l Conference on Cyber, Physical and Social Computing (CPSCom), IEEE, 2010, pp. 179–188.

[31] Y. Ding, X. Qin, L. Liu, T. Wang, Energy efficient scheduling of virtual machines in cloud with deadline constraint, Future Gener. Comput. Syst. (2015).

[32] A. Berl, E. Gelenbe, M. Di Girolamo, G. Giuliani, H. De Meer, M.Q. Dang, K. Pentikousis, Energy-efficient cloud computing, Comput. J. 53 (7) (2010) 1045–1051.

[33] A. Greenberg, J. Hamilton, D.A. Maltz, P. Patel, The cost of a cloud: research problems in data center networks, ACM SIGCOMM Comput. Commun. Rev. 39 (1) (2008) 68–73.

[34] A. Khosravi, S.K. Garg, R. Buyya, Energy and carbon-efficient placement of virtual machines in distributed cloud data centers, in: Euro-Par 2013 Parallel Processing, Springer, Berlin, Heidelberg, 2013, pp. 317–328.

[35] L.A. Barroso, U. Hölzle, The case for energy-proportional computing, IEEE Comput. 40 (12) (2007) 33–37.

[36] A. Verma, P. Ahuja, A. Neogi, pMapper: power and migration cost aware application placement in virtualized systems, in: Middleware 2008, Springer, Leuven, Belgium, 2008, pp. 243–264.

[37] C. Calwell, T. Reeder, Power supplies: a hidden opportunity for energy savings, Technical Report, Natural Resources Defense Council, San Francisco, CA, 2002.

[38] D. Meisner, B.T. Gold, T.F. Wenisch, PowerNap: eliminating server idle power, ACM SIGPLAN Not. 44 (2009) 205–216.

[39] R. Buyya, C.S. Yeo, S. Venugopal, J. Broberg, I. Brandic, Cloud computing and emerging IT platforms: vision, hype, and reality for delivering computing as the 5th utility, Future Gener. Comput. Syst. 25 (6) (2009) 599–616.

[40] L. Schubert, K.G. Jeffery, B. Neidecker-Lutz, The Future of Cloud Computing: Opportunities for European Cloud Computing Beyond 2010. Expert Group Report, European Commission, Information Society and Media, Brussels, Belgium, 2010.

[41] R.J. Adair, A virtual machine system for the 360/40, International Business Machines Corporation, Cambridge Scientific Center, 1966.

[42] R.P. Goldberg, Survey of virtual machine research, Computer 7 (6) (1974) 34–45.

[43] S. Loveland, E.M. Dow, F. LeFevre, D. Beyer, P.F. Chan, Leveraging virtualization to optimize high-availability system configurations, IBM Syst. J. 47 (2008) 591–604.

[44] C. Clark, K. Fraser, S. Hand, J.G. Hansen, E. Jul, C. Limpach, I. Pratt, A. Warfield, Live migration of virtual machines, in: Proceedings of the 2nd Conference on Symposium on Networked Systems Design & Implementation, vol. 2, USENIX Association, 2005, pp. 273–286.

[45] B. Cully, G. Lefebvre, D. Meyer, M. Feeley, N. Hutchinson, A. Warfield, Remus: high availability via asynchronous virtual machine replication, in: Proceedings of the 5th USENIX Symposium on Networked Systems Design and Implementation, San Francisco, 2008, pp. 161–174.

[46] Y. Tamura, Kemari: Virtual Machine Synchronization for Fault Tolerance Using DomT, Xen Summit, 2008.

[47] L. Wang, Z. Kalbarczyk, R.K. Iyer, A. Iyengar, Checkpointing virtual machines against transient errors, in: 2010 IEEE 16th International On-Line Testing Symposium (IOLTS), IEEE, 2010, pp. 97–102.

[48] H.A. Lagar-Cavilla, J.A. Whitney, A.M. Scannell, P. Patchin, S.M. Rumble, E. De Lara, M. Brudno, M. Satyanarayanan, SnowFlock: rapid virtual machine cloning for cloud computing, in: Proceedings of the 4th ACM European Conference on Computer Systems, ACM, 2009, pp. 1–12.

[49] K.-Y. Hou, M. Uysal, A. Merchant, K.G. Shin, S. Singhal, HydraVM: Low-Cost, Transparent High Availability for Virtual Machines, Technical Report, HP Laboratories, 2011.

[50] L. Cherkasova, D. Gupta, A. Vahdat, Comparison of the three CPU schedulers in Xen, SIGMETRICS Perform. Eval. Rev. 35 (2) (2007) 42–51.

[51] D. Schanzenbach, H. Casanova, Accuracy and responsiveness of CPU sharing using Xens cap values, Technical Report ICS2008-05-01, University of Hawaii at Manoa Department of Information and Computer Sciences, 2008.

[52] X. Xu, P. Shan, J. Wan, Y. Jiang, Performance evaluation of the CPU scheduler in XEN, in: International Symposium on Information Science and Engineering, 2008, ISISE'08, vol. 2, IEEE, 2008, pp. 68–72.

[53] Z. Shen, S. Subbiah, X. Gu, J. Wilkes, Cloudscale: elastic resource scaling for multi-tenant cloud systems, in: Proceedings of the 2nd ACM Symposium on Cloud Computing, ACM, 2011, p. 5.

[54] D. Armstrong, K. Djemame, Towards quality of service in the cloud, in: Proceedings of the 25th UK Performance Engineering Workshop, 2009, pp. 1–15.

[55] V. Stantchev, C. Schröpfer, Negotiating and enforcing QoS and SLAs in grid and cloud computing, in: Advances in Grid and Pervasive Computing, Springer, Berlin, Heidelberg, 2009, pp. 25–35.

[56] P. Patel, A.H. Ranabahu, A.P. Sheth, Service level agreement in cloud computing, in: Proceedings of the Cloud Workshops at OOPSLA09, 2009, pp. 25–29.

[57] A. Keller, H. Ludwig, The WSLA framework: specifying and monitoring service level agreements for web services, J. Netw. Syst. Manag. 11 (1) (2003) 57–81.

[58] M. Keller, D. Meister, A. Brinkmann, C. Terboven, C. Bischof, eScience cloud infrastructure, in: 2011 37th EUROMICRO Conference on Software Engineering and Advanced Applications (SEAA), IEEE, 2011, pp. 188–195.

[59] C. Vecchiola, S. Pandey, R. Buyya, High-performance cloud computing: a view of scientific applications, in: 2009 10th International Symposium on Pervasive Systems, Algorithms, and Networks (ISPAN), IEEE, 2009, pp. 4–16.

[60] S.K. Garg, A.N. Toosi, S.K. Gopalaiyengar, R. Buyya, SLA-based virtual machine management for heterogeneous workloads in a cloud datacenter, J. Netw. Comput. Appl. 45 (2014) 108–120.

[61] A. Beloglazov, J. Abawajy, R. Buyya, Energy-aware resource allocation heuristics for efficient management of data centers for cloud computing, Future Gener. Comput. Syst. 28 (5) (2012) 755–768.

[62] G. Dhiman, G. Marchetti, T. Rosing, vGreen: a system for energy efficient computing in virtualized environments, in: Proceedings of the 14th ACM/IEEE International Symposium on Low Power Electronics and Design, ACM, 2009, pp. 243–248.

[63] Y.C. Lee, A.Y. Zomaya, Energy efficient utilization of resources in cloud computing systems, J. Supercomput. 60 (2) (2012) 268–280.

[64] C. Mastroianni, M. Meo, G. Papuzzo, Probabilistic consolidation of virtual machines in self-organizing cloud data centers, IEEE Trans. Cloud Comput. 1 (2) (2013) 215–228.

[65] Q. Zhu, J. Zhu, G. Agrawal, Power-aware consolidation of scientific workflows in virtualized environments, in: Proceedings of the 2010 ACM/IEEE International Conference for High Performance Computing, Networking, Storage and Analysis, IEEE Computer Society, 2010, pp. 1–12.

[66] I.S. Moreno, P. Garraghan, P. Townend, J. Xu, Analysis, modeling and simulation of workload patterns in a large-scale utility cloud, IEEE Trans. Cloud Comput. 2 (2) (2014) 208–221.

[67] P. Garraghan, I.S. Moreno, P. Townend, J. Xu, An analysis of failure-related energy waste in a large-scale cloud environment, IEEE Trans. Emerg. Top. Comput. 2 (2) (2014) 166–180.

[68] C. Reiss, A. Tumanov, G.R. Ganger, R.H. Katz, M.A. Kozuch, Towards understanding heterogeneous clouds at scale: Google trace analysis, Technical Report, Intel Science and Technology Center for Cloud Computing, 2012.

[69] Google, Google cluster data v2. http://code.google.com/p/googleclusterdata/wiki/ClusterData2011_1, 2011 (accessed 25.03.13).

[70] R.N. Calheiros, R. Ranjan, A. Beloglazov, C.A. De Rose, R. Buyya, CloudSim: a toolkit for modeling and simulation of cloud computing environments and evaluation of resource provisioning algorithms, Softw. Pract. Exp. 41 (1) (2011) 23–50.

[71] I. Philip, Software failures and the road to a petaflop machine, in: HPCRI: 1st Workshop on High Performance Computing Reliability Issues, in Proceedings of the 11th International Symposium on High Performance Computer Architecture (HPCA-11), 2005.

[72] B. Schroeder, G.A. Gibson, Understanding failures in petascale computers, J. Phys. Conf. Ser. 78 (2007) 012022.

[73] M.L. Shooman, Reliability of Computer Systems and Networks: Fault Tolerance, Analysis, and Design, John Wiley & Sons, Inc., New York, 2002, ISBN 0471293423.

[74] D. Kececioglu, Reliability Engineering Handbook, vol. 1, DEStech Publications, Lancaster, PA, 2002.

[75] A. Avizienis, J.-C. Laprie, B. Randell, C. Landwehr, Basic concepts and taxonomy of dependable and secure computing, IEEE Trans. Dependable Secure Comput. 1 (1) (2004) 11–33.

[76] C. Dabrowski, Reliability in grid computing systems, Concurr. Comput. 21 (8) (2009) 927–959.

[77] I.P. Egwutuoha, D. Levy, B. Selic, S. Chen, A survey of fault tolerance mechanisms and checkpoint/restart implementations for high performance computing systems, J. Supercomput. 65 (3) (2013) 1302–1326.

[78] I. Koren, C.M. Krishna, Fault-Tolerant Systems, Morgan Kaufmann, Burlington, MA, 2010.

[79] M. Young, M. Accetta, R. Baron, W. Bolosky, D. Golub, R. Rashid, A. Tevanian, Mach: a new kernel foundation for UNIX development, in: Proceedings of the 1986 Summer USENIX Conference, 1986, pp. 93–113.

[80] C. Minyard, A gentle introduction with OpenIPMI, MontaVista Software. URL http://openipmi.sourceforge.net/IPMI.pdf, 2006.

[81] Hardware monitoring by lm-sensors, lm_sensors—Linux hardware monitoring. http://www.lm-sensors.org/, 2010 (accessed 24.12.14).

[82] S. Fu, C.-Z. Xu, Exploring event correlation for failure prediction in coalitions of clusters, in: Proceedings of the 2007 ACM/IEEE Conference on Supercomputing, 2007, SC'07, IEEE, 2007, pp. 1–12.

[83] J.W. Mickens, B.D. Noble, Exploiting availability prediction in distributed systems, in: Proceedings of the 3rd USENIX/ACM Symposium on Networked Systems Design and Implementation (NSDI), San Jose, CA, 2006, pp. 73–86.

[84] N. Yigitbasi, M. Gallet, D. Kondo, A. Iosup, D. Epema, Analysis and modeling of time-correlated failures in large-scale distributed systems, in: 2010 11th IEEE/ACM International Conference on Grid Computing (GRID), IEEE, 2010, pp. 65–72.

[85] S. Fu, C.-Z. Xu, Quantifying temporal and spatial correlation of failure events for proactive management, in: 26th IEEE International Symposium on Reliable Distributed Systems, 2007, SRDS 2007, IEEE, 2007, pp. 175–184.

[86] F. Salfner, M. Lenk, M. Malek, A survey of online failure prediction methods, ACM Comput. Surv. 42 (3) (2010) 10.

[87] C. Engelmann, G.R. Vallee, T. Naughton, S.L. Scott, Proactive fault tolerance using preemptive migration, in: 17th Euromicro International Conference on Parallel, Distributed and Network-Based Processing, 2009, IEEE, 2009, pp. 252–257.

[88] G. Casale, S. Kraft, D. Krishnamurthy, A model of storage I/O performance interference in virtualized systems, in: 2011 31st International Conference on Distributed Computing Systems Workshops (ICDCSW), IEEE, 2011, pp. 34–39.

[89] Y. Mei, L. Liu, X. Pu, S. Sivathanu, Performance measurements and analysis of network I/O applications in virtualized cloud, in: 2010 IEEE 3rd International Conference on Cloud Computing (CLOUD), IEEE, 2010, pp. 59–66.

[90] G. Somani, S. Chaudhary, Application performance isolation in virtualization, in: IEEE International Conference on Cloud Computing, 2009, CLOUD'09, IEEE, 2009, pp. 41–48.

[91] Q. Zhu, T. Tung, A performance interference model for managing consolidated workloads in QoS-aware clouds, in: 2012 IEEE 5th International Conference on Cloud Computing (CLOUD), IEEE, 2012, pp. 170–179.

[92] S. Blagodurov, S. Zhuravlev, A. Fedorova, Contention-aware scheduling on multicore systems, ACM Trans. Comput. Syst. 28 (4) (2010) 8.

[93] Q. Huang, P.P. Lee, An experimental study of cascading performance interference in a virtualized environment, ACM SIGMETRICS Perform. Eval. Rev. 40 (4) (2013) 43–52.

[94] X. Pu, L. Liu, Y. Mei, S. Sivathanu, Y. Koh, C. Pu, Y. Cao, Who is your neighbor: net I/O performance interference in virtualized clouds, IEEE Trans. Serv. Comput. 6 (3) (2013) 314–329.

[95] S. Govindan, J. Liu, A. Kansal, A. Sivasubramaniam, Cuanta: quantifying effects of shared on-chip resource interference for consolidated virtual machines, in: Proceedings of the 2nd ACM Symposium on Cloud Computing, ACM, 2011, p. 22.

[96] J. Mars, L. Tang, K. Skadron, M.L. Soffa, R. Hundt, Increasing utilization in modern warehouse-scale computers using bubble-up, IEEE Micro 32 (3) (2012) 88–99.

[97] T. Dwyer, A. Fedorova, S. Blagodurov, M. Roth, F. Gaud, J. Pei, A practical method for estimating performance degradation on multicore processors, and its application to HPC workloads, in: Proceedings of the International Conference on High Performance Computing, Networking, Storage and Analysis, IEEE Computer Society Press, 2012, p. 83.

[98] X. Pu, L. Liu, Y. Mei, S. Sivathanu, Y. Koh, C. Pu, Understanding performance interference of I/O workload in virtualized cloud environments, in: 2010 IEEE 3rd International Conference on Cloud Computing (CLOUD), IEEE, 2010, pp. 51–58.

[99] G. Somani, S. Chaudhary, Performance isolation and scheduler behavior, in: 2010 1st International Conference on Parallel Distributed and Grid Computing (PDGC), IEEE, 2010, pp. 272–277.

[100] Y. Koh, R. Knauerhase, P. Brett, M. Bowman, Z. Wen, C. Pu, An analysis of performance interference effects in virtual environments, in: IEEE International Symposium on Performance Analysis of Systems & Software, 2007, ISPASS 2007, IEEE, 2007, pp. 200–209.

[101] S.-H. Lim, J.-S. Huh, Y. Kim, G.M. Shipman, C.R. Das, D-factor: a quantitative model of application slow-down in multi-resource shared systems, ACM SIGMETRICS Perform. Eval. Rev. 40 (1) (2012) 271–282.

[102] I.S. Moreno, R. Yang, J. Xu, T. Wo, Improved energy-efficiency in cloud datacenters with interference-aware virtual machine placement, in: 2013 IEEE Eleventh International Symposium on Autonomous Decentralized Systems (ISADS), IEEE, 2013, pp. 1–8.

[103] F. Hermenier, N. Loriant, J.-M. Menaud, Power management in grid computing with Xen, in: Frontiers of High Performance Computing and Networking—ISPA 2006 Workshops, Springer, 2006, pp. 407–416.

[104] L. Lefèvre, A.-C. Orgerie, Designing and evaluating an energy efficient cloud, J. Supercomput. 51 (3) (2010) 352–373.

[105] E. Falkenauer, A. Delchambre, A genetic algorithm for bin packing and line balancing, in: Proceedings of the 1992 IEEE International Conference on Robotics and Automation, IEEE, 1992, pp. 1186–1192.

[106] J. Xu, J. Fortes, A multi-objective approach to virtual machine management in datacenters, in: Proceedings of the 8th ACM International Conference on Autonomic Computing, ACM, 2011, pp. 225–234.

[107] J.-L. Deneubourg, S. Goss, N. Franks, A. Sendova-Franks, C. Detrain, L. Chrétien, The dynamics of collective sorting robot-like ants and ant-like robots, in: From Animals to Animats: Proceedings of the First International Conference on Simulation of Adaptive Behavior, 1991, pp. 356–363.

[108] J.A. Nelder, R. Mead, A simplex method for function minimization, Comput. J. 7 (4) (1965) 308–313.

[109] A.M. Sampaio, J.G. Barbosa, Dynamic power- and failure-aware cloud resources allocation for sets of independent tasks, in: 2013 IEEE International Conference on Cloud Engineering (IC2E), IEEE, 2013, pp. 1–10.

[110] A.M. Sampaio, J.G. Barbosa, Towards high-available and energy-efficient virtual computing environments in the cloud, Future Gener. Comput. Syst. 40 (2014) 30–43.

[111] A. Verma, G. Dasgupta, T.K. Nayak, P. De, R. Kothari, Server workload analysis for power minimization using consolidation, in: Proceedings of the 2009 Conference on USENIX Annual Technical Conference, USENIX Association, 2009, p. 28.

[112] S. Fu, Failure-aware resource management for high-availability computing clusters with distributed virtual machines, J. Parallel Distrib. Comput. 70 (4) (2010) 384–393.

[113] H. Arabnejad, J. Barbosa, List scheduling algorithm for heterogeneous systems by an optimistic cost table, IEEE Trans. Parallel Distrib. Syst. 25 (2014) 682–694.

[114] I.I. Yusuf, H.W. Schmidt, I.D. Peake, Evaluating recovery aware components for grid reliability, in: Proceedings of the 7th Joint Meeting of the European Software Engineering Conference and the ACM SIGSOFT Symposium on the Foundations of Software Engineering, ACM, 2009, pp. 277–280.

[115] I.I. Yusuf, H.W. Schmidt, I.D. Peake, Architecture-based fault tolerance support for grid applications, in: Proceedings of the Joint ACM SIGSOFT Conference—QoSA and ACM SIGSOFT Symposium—ISARCS on Quality of Software Architectures—QoSA and Architecting Critical Systems—ISARCS, ACM, 2011, pp. 177–182.

[116] I.I. Yusuf, H.W. Schmidt, Parameterised architectural patterns for providing cloud service fault tolerance with accurate costings, in: Proceedings of the 16th International ACM Sigsoft Symposium on Component-Based Software Engineering, ACM, 2013, pp. 121–130.

[117] A. Butoi, A. Stan, G.C. Silaghi, Autonomous management of virtual machine failures in IaaS using fault tree analysis, in: Economics of Grids, Clouds, Systems, and Services, Springer International Publishing, Switzerland, 2014, pp. 206–221.

[118] Y. Li, Z. Lan, Exploit failure prediction for adaptive fault-tolerance in cluster computing, in: Sixth IEEE International Symposium on Cluster Computing and the Grid, 2006, CCGRID 06, vol. 1, IEEE, 2006, pp. 531–538.

[119] A.B. Nagarajan, F. Mueller, C. Engelmann, S.L. Scott, Proactive fault tolerance for HPC with Xen virtualization, in: Proceedings of the 21st Annual International Conference on Supercomputing, ACM, 2007, pp. 23–32.

[120] A.J. Oliner, R.K. Sahoo, J.E. Moreira, M. Gupta, A. Sivasubramaniam, Fault-aware job scheduling for BlueGene/L systems, in: Proceedings of the 18th International Parallel and Distributed Processing Symposium, 2004, IEEE, 2004, p. 64.

[121] A. Polze, P. Troger, F. Salfner, Timely virtual machine migration for pro-active fault tolerance, in: 2011 14th IEEE International Symposium on Object/Component/Service-Oriented Real-Time Distributed Computing Workshops (ISORCW), IEEE, 2011, pp. 234–243.

[122] H. Salami, H. Saadatfar, F.R. Fard, S.K. Shekofteh, H. Deldari, Improving cluster computing performance based on job futurity prediction, in: 2010 3rd International Conference on Advanced Computer Theory and Engineering (ICACTE), vol. 6, IEEE, 2010, p. V6–303.

[123] S.-g. Kim, H. Eom, H.Y. Yeom, Virtual machine scheduling for multicores considering effects of shared on-chip last level cache interference, in: 2012 International Green Computing Conference (IGCC), IEEE, 2012, pp. 1–6.

[124] R.C. Chiang, H.H. Huang, TRACON: interference-aware scheduling for data-intensive applications in virtualized environments, in: Proceedings of 2011 International Conference for High Performance Computing, Networking, Storage and Analysis, ACM, 2011, p. 47.

[125] R. Nathuji, A. Kansal, A. Ghaffarkhah, Q-clouds: managing performance interference effects for QoS-aware clouds, in: Proceedings of the 5th European Conference on Computer Systems, ACM, 2010, pp. 237–250.

[126] A.M. Sampaio, J.G. Barbosa, R. Prodan, PIASA: a power and interference aware resource management strategy for heterogeneous workloads in cloud data centers, Simul. Model. Pract. Theory 57 (2015) 142–160.

[127] A.M. Sampaio, J.G. Barbosa, Optimizing energy-efficiency in high-available scientific cloud environments, in: 2013 Third International Conference on Cloud and Green Computing (CGC), IEEE, 2013, pp. 76–83.

[128] SPECpower, First quarter 2014 SPECpower_ssj2008 results. http://www.spec.org/power_ssj2008/results/res2014q1/, 2014 (accessed 15.01.15).

[129] W. Zhao, P. Melliar-Smith, L.E. Moser, Fault tolerance middleware for cloud computing, in: 2010 IEEE 3rd International Conference on Cloud Computing (CLOUD), IEEE, 2010, pp. 67–74.

ABOUT THE AUTHORS

Altino M. Sampaio received his BSc degree in Electrical and Computers Engineering from the Faculty of Engineering of the University of Porto (FEUP), Porto, Portugal, in 2002, and the PhD degree in Informatics Engineering from FEUP, Portugal, in 2015. Since 2008 he serves as a teacher assistant in several computer science courses at Instituto Politécnico do Porto, Porto, Portugal. His research interests are in the fields of Cloud and distributed computing systems, especially resources scheduling, considering fault tolerance and energy optimization in Cloud Computing.

Jorge G. Barbosa received the BSc degree in Electrical and Computer Engineering from Faculty of Engineering of the University of Porto (FEUP), Portugal, the MSc in Digital Systems from University of Manchester Institute of Science and Technology, England, in 1993, and the PhD in Electrical and Computer Engineering from FEUP, Portugal, in 2001. Since 2001 he is an Assistant Professor at FEUP. His research interests are related to parallel and distributed computing, heterogeneous computing, scheduling in heterogeneous environments, cloud computing and biomedical engineering.

CHAPTER FOUR

Achieving Energy Efficiency in Datacenters by Virtual Machine Sizing, Replication, and Placement

Hadi Goudarzi*, Massoud Pedram†
*Qualcomm Incorporated, San Diego, California, USA
†Department of Electrical Engineering, University of Southern California, Los Angeles, California, USA

Contents

Abstract

Monitoring and analyzing virtual machines (VMs) in a datacenter enables a cloud provider to place them on fewer physical machines with negligible performance penalty, a process which is known as the VM consolidation. Such a consolidation allows the cloud provider to take advantage of dissimilar workloads to reduce the number of ON servers,

Advances in Computers, Volume 100
ISSN 0065-2458
http://dx.doi.org/10.1016/bs.adcom.2015.11.001

161

and thereby, reduce the datacenter energy consumption. Moreover, placing multiple copies of VMs on different servers and distributing the incoming requests among them can reduce the resource requirement for each copy and help the cloud provider do more aggressive consolidation. This chapter begins by a substantial review of various approaches for consolidation, resource management, and power control in datacenters. It continues by presenting a dynamic programming-based algorithm for creating multiple copies of a VM without degrading performance and doing VM consolidation for the purpose of datacenter energy minimization. A side benefit of the consolidation is to improve reliability of the services provided by the datacenter to its clients. Using the proposed algorithm, it is shown that more than 20% energy saving can be achieved compared to the previous work.

1. INTRODUCTION

Demand for computing power has been increasing due to the penetration of information technologies in our daily interactions with the world both at personal and public levels, encompassing business, commerce, education, manufacturing, and communication services. At personal level, the wide-scale presence of online banking, e-commerce, SaaS (Software as a Service), social networking, and so on produce workloads of great diversity and enormous scale. At the same time, computing and information processing requirements of various public organizations and private corporations have also been increasing rapidly. Examples include digital services and functions required by the various industrial sectors, ranging from manufacturing to housing, from transportation to banking. Such a dramatic increase in the computing demand requires a scalable and dependable IT infrastructure comprising of servers, storage, network bandwidth, physical infrastructure, electrical grid, IT personnel and billions of dollars in capital expenditure, and operational cost to name a few.

Datacenters are the backbone of today's IT infrastructure. The reach of datacenters spans a broad range of application areas from energy production and distribution, complex weather modeling and prediction, manufacturing, transportation, entertainment, and even social networking. There is a critical need to continue to improve efficiency in all these sectors by accelerated use of computing technologies, which inevitably requires increasing the size and scope of datacenters. However, datacenters themselves are now faced with a major impediment of power consumption. Some reports such as [1,2] estimate that the datacenter electricity demand in 2012 was around 31 GW globally, which is equivalent to the electricity demand of around

23 million homes. These reports also predict fast growth rate for electrical energy consumption in datacenters. Power consumption of datacenters will soon match or exceed many other energy-intensive industries such as air transportation.

Apart from the total energy consumption, another critical component is the peak power; according to an EPA report [3], the peak load on the power grid from datacenters is estimated to be approximately 7 GW in 2006 in the United States, equivalent to the output of about 15 base-load power plants. This load is increasing as shipments of high-end servers used in datacenters (e.g., blade servers) are increasing at a 20–30% CAGR.

System-wide power management is a huge challenge in datacenters. First, restrictions on availability of power and large power consumption of the IT equipment make the problem of datacenter power management a very difficult one to cope with. Second, the physical infrastructure (e.g., the power backup and distribution system and the computer room air conditioning systems) tends to account for up to one-third of total datacenter power and capital costs [4–6]. Third, the peak instantaneous power consumption must be controlled. The reason for capping power dissipation in the datacenters is the capacity limitation of the power delivery network in the datacenter facility. Fourth, power budgets in datacenters exist in different granularities: datacenter, cluster, rack, or even servers. A difficulty in the power capping is the distributed nature of power consumption in the datacenter. For example, if there is a power budget for a rack in the datacenter, the problem is how to allocate this budget to different servers and how to control this budget in a distributed fashion. Finally, another goal is to reduce the total power consumption. A big portion of the datacenter operational cost is the cost of electrical energy purchased from the utility companies. A trade-off exists between power consumption and performance of the system and the power manager should consider this trade-off carefully. For example, if the supply voltage level and clock frequency of a CPU are reduced, the average power consumption (and even energy needed to execute a given task) is reduced, but the total computation time is increased.

Low utilization of servers in a datacenter is one of the biggest factors in low power efficiency of the datacenter. The most important reason behind having the best energy efficiency at 100% load in servers is the energy non-proportional behavior of the servers [7]. This means that servers with idle status consume a big portion of their peak power consumption. The fact that most of the times, servers are utilized with between 10% and 50% of their peak load and discrete frequent idle times of servers [8] amplify this issue in the datacenters. This fact motivates the design of energy-proportional servers [4] to minimize the overall power consumption. However, due to the nonenergy-proportional nature of the current servers, it is prudent from an energy efficiency viewpoint to have as few servers as possible turned on with each active server being highly utilized. In order to decrease the number of active servers, sharing a physical server between several applications is necessary. Virtualization technology creates this opportunity.

Virtualization technology creates an application-hosting environment that provides independence between applications that share a physical machine together [9]. Nowadays, computing systems rely heavily on this technology. Virtualization technology provides a new way to improve the power efficiency of the datacenters: consolidation. Consolidation means assigning more than one virtual machine (VM) to a physical server. As a result, some of the servers can be turned off and power consumption of the computing system decreases. Again the technique involves performance-power trade-off. More precisely, if workloads are consolidated on servers, performance of the consolidated VMs may decrease because of physical resource contention (CPU, memory, I/O bandwidth) but the power efficiency will improve because fewer servers will be used to service the VMs.

In order to determine the amount of the resources that needs to be allocated to each VM, some performance target needs to be defined for each VM. The IT infrastructure provided by the datacenter owners/operators must meet various service level agreements (SLAs) established with the clients. The SLAs may be resource related (e.g., amount of computing power, memory/storage space, network bandwidth), performance related (e.g., service time or throughput), or even quality of service (QoS) related (24-7 availability, data security, percentage of dropped requests). SLA constraints can be used to determine the limit (minimum and maximum) on the resource requirement of each VM to be able to satisfy the required performance target. On the other hand, in order to minimize the operational cost of the datacenter, energy cost also needs to be considered to decide about optimal resource assignment to VMs.

The scale of the resource management problem in datacenters is very big because a datacenter comprises of thousands to tens of thousands of server machines, working in tandem to provide services to hundreds of thousands clients at the same time, see for example, Refs. [10,11]. In such a large computing system, energy efficiency can be maximized through system-wide resource allocation and VM consolidation. This is in spite of nonenergy-proportional characteristics of current server machines [7].

Resource management solution affects the operational cost and admission control policy in the cloud computing system. Resource management in datacenter is usually handled by three types of resource manager: resource arbiter, power manager, and thermal managers. Resource arbiter or VM manager decides about VM to server assignment and migration and resource allocation. Power manager controls the average and peak power in a distributed or centralized fashion in datacenters and thermal manager keeps the hardware temperature below certain critical point and minimizes the power consumption of the cooling system. In this chapter, a review of the most important work in the area of the resource arbiter and power manager is presented. Moreover, a novel approach to minimize the energy cost of datacenter by increasing the VM consolidation opportunity using VM replication is proposed.

Generating multiple copies of a VM and placing them on different servers is one of the basic ways to increase the service reliability. In this approach, only the original copy of the VM handles the requests and the other copies are idle. In this chapter, we propose to exploit all of these copies for servicing the requests. In this scenario, resource provided for each copy of the VM should satisfy SLA requirements and the set of distributed VMs should be able to service all of the incoming requests. For this reason, memory bandwidth (bandwidth) provided for each copy of the VM should be the same as that of the original VM whereas the total CPU cycles provided for these VMs should be greater or equal to the provided CPU cycles for the original VM. Using this approach and an effective VM placement algorithm, which determines the number of VMs and place them on physical machines, the energy cost of the system can be reduced by 20%.

The proposed VM replication and placement algorithm is based on the dynamic programming (DP) and local search methods. The DP method determines the number of copies for each VM and places them on servers and the local search tries to minimize the energy cost by turning off the under-utilized servers.

The rest of this chapter is organized as follows. Related work is presented in Section 2. The system model and problem formulation are presented in Sections 3 and 4. The proposed algorithm is presented in Section 5. The simulation results are presented in Section 6, and the conclusions and future work directions are presented in Section 7.

2. RELATED WORK IN DATACENTER POWER AND RESOURCE MANAGEMENT

A datacenter resource management system is comprised of three main components: resource arbiter, power manager, and temperature manager. An exemplary architecture for the datacenter resource management system with emphasis on the resource arbiter is depicted in Fig. 1.

In this architecture, the resource arbiter handles the application placement into resources and application migration. In this chapter, the terms resource arbiter and resource manager are used interchangeably.

To assign applications or VMs to resources in a datacenter, one must monitor the resource availability and performance state of the physical servers in the datacenter. In addition, the resource arbiter must interact with

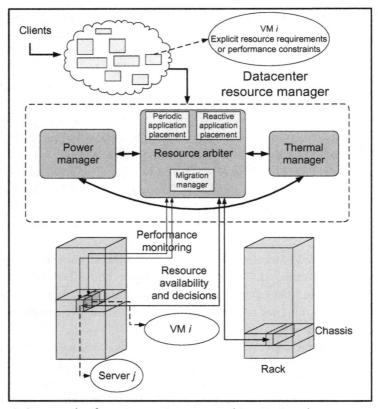

Figure 1 An example of resource management architecture in a datacenter.

the power and thermal managers. For example, if the power manager has limited the maximum power consumption of server, the resource arbiter should consider this limitation when assigning a new VM to the server. On the other hand, the power manager tries to minimize the average power consumption in each server considering the performance constraints of VMs assigned to servers as determined by the resource arbiter. Similarly, the resource arbiter must use the information provided by the thermal manager to decrease the workload of hot servers. At the same time, the thermal manager tries to control the temperature of active servers while accepting VM to server assignments made by the resource arbiter and meeting the per-VM performance constraints set forth by the arbiter.

In this section, a review of the important approaches and techniques for design and optimization of resource arbiter and power managers in datacenters is presented. Thermal managers are out of the scope of this chapter. The review is by no means comprehensive, but aims to present some key approaches and results.

2.1 Resource Arbiter in Datacenters

Several versions of the resource management problem in datacenters have been investigated in the literature. Some of the prior work focuses on maximizing the number of served tasks in a datacenter (or total revenue for the datacenter operator) without considering the energy cost. Example references are [12,13], where the authors present heuristic solutions based on network flow optimization to find a revenue maximizing solution for a scenario in which the total resource requirement of tasks is more than the total resource capacity in the datacenter. The resource assignment problem for tasks with fixed memory, disk, and processing requirements is tackled in [14], where the authors describe an approximation algorithm for solving the problem of maximizing the number of tasks serviced in the datacenter.

Another version of the resource management problem is focused on minimizing the total electrical energy cost. Key considerations are to service all incoming tasks while satisfying specified performance guarantees for each task. A classic example of this approach is the work of Chase *et al.* [15] who present a resource assignment solution in a hosting datacenter with the objective of minimizing the energy consumption while responding to power supply disruptions and/or thermal events. In this paper, economics-based approaches are used to manage the resource allocation in a system with shared resources in which clients bid for resources as a function of delivered performance.

Yet, another version of the resource management problem considers the server and cooling power consumptions during the resource assignment problem. A representative of approaches to solving this problem is Ref. [16], in which Pakbaznia *et al.* present a solution for concurrent task assignment and VM consolidation in regular period called epochs. More precisely, workload prediction is used to determine the resource requirements (and hence the number of ON servers) for all incoming tasks for the epoch. Next, considering the current datacenter temperature map and using an analytical model for predicting the future temperature map as a function of the server power dissipations, locations of the ON servers for the next epoch are determined, and tasks are assigned to the ON servers so that the total datacenter power consumption is minimized.

Considering the effect of consolidation on the performance of servers is the key to reducing the total power consumption in a datacenter without creating unacceptable performance degradations. For example, Srikantaiah *et al.* [17] present an energy-aware resource assignment technique based on an experimental study of the performance, energy usage, and resource utilization of the servers while employing VM consolidation. In particular, two dimensions for server resources are considered in this paper: disk and CPU. Effects of the consolidation on performance degradation and energy consumption per transaction are quantified. The authors recommend applying consolidation so as not to over-commit servers in any resource dimension. The problem of application placement into a minimum number of ON servers, which is equivalent to the well-known bin-packing problem, is discussed, and a greedy algorithm for solving it is described.

Correlation of resource utilization patterns among VMs is an important factor when VM consolidation decision is being made [18]. Assigning highly (positive) correlated VMs in terms of resource usage, will increase the chance of VM migrations that is needed to avoid SLA violation. On the other hand, consolidating VMs that have less correlation in terms of their resource usage pattern results in more packed servers and lower power consumption [19]. Practical experiments of this theory are presented in [20], which suggests that the interferences between consolidated VMs in terms of (CPU/memory/networking) resource usage can cause the resource utilization to be lower or higher than the summation of the resource usage for VMs assigned to the server and needs to be considered to avoid performance degradation and SLA violations.

Resource usage of a VM in server can interfere with other VMs placed on that server. Moreover, VMs can have uneven resource utilization along different dimensions. These issues need to be considered in VM consolidation decisions. For example, the effect of uneven resource utilization along different dimensions (e.g., CPU, memory, and I/O) by different VMs and the question of how to improve datacenter energy efficiency by increasing the resource utilization in different resource dimensions are investigated by Xiao et al. [21]. An approach to resolve the interference between VMs placed on the same physical machine is presented in [22].

A technique to maximize the utilization of the active server while creating more idle servers that can subsequently be turned off is to migrate VMs from a server with a low utilization factor to another server. A good example of considering server power consumption and VM migration cost in the resource assignment problem is Ref. [23], which presents power and migration cost-aware application placement in a virtualized datacenter. For this problem, each VM has fixed and known resource requirements based on the specified SLA. An elaborate architecture called pMapper and an effective VM placement algorithm to solve the assignment problem are key components of the proposed solution. More precisely, various actions in pMapper algorithm are classified as: (i) soft actions like VM re-sizing, (ii) hard actions such as dynamic voltage frequency scaling (DVFS), and (iii) VM consolidation actions. These actions are implemented by different parts of the implemented middleware. There is a resource arbiter, which has a global view of the applications and their SLAs and issues soft action commands. A power manager issues hard action commands whereas a migration manager triggers consolidation decisions in coordination with a virtualization manager. These managers communicate with an arbitrator as the global decision maker to set the VM sizes and find a good application placement based on the inputs of different managers. Any revenue losses due to performance degradation caused by VM migration are calculated considering the given SLAs and used to set the migration costs of VMs. To optimally place VMs onto servers, the authors rely on a power efficiency metric to statically rank the servers independent of the applications running on them. This is because creating a dynamic ranking model for all mixes of all applications on all servers is infeasible. A heuristic based on the first-fit decreasing (FFD) bin-packing algorithm [24] is presented to place the applications on servers starting with the most power-efficient server. Different versions of the FFD solution are proposed in a number of previous works including [25,26] to decide about VM assignment and consolidation.

The problem of resource allocation is more challenging in case of having clients with SLA contracts for a datacenter owner who wants to maximize its profit by reducing the SLA violations and decrease the operational cost [27]. Many researchers in different fields have addressed the problem of SLA-driven resource assignment. Some of the previous work have considered probabilistic SLA constraints with violation penalty, e.g., Refs. [28,29]. Other work has relied on utility function-based SLA [26,30–32]. In [33], an SLA with soft constraint on average response time is considered for multitier applications to solve the resource assignment problem. To determine and adjust the amount of resource allocated to VMs to satisfy SLA constraints, approaches based on the reinforcement learning [34] and look-ahead control theory [35] have also been proposed. SLA contracts with guarantee on response time and/or penalties paid for violating the stipulated response time constraint are considered in [36]. In this paper, a resource management system is presented that determines the amount of resource that needs to be allocated to VMs based on SLA contracts and energy cost and subsequently assigns VMs to servers so as to reduce the operational cost of datacenter.

Due to big number of VMs and servers in a datacenter, an important factor in designing VM management solution is to make it as scalable as possible. Different works in the literature tackles this problem. Feller et al. [37] present a fully decentralized VM control solution to make the VM consolidation decisions. The proposed solution is based on peer-to-peer communication between physical servers to decide about assignment of new VM to a server and migration of VMs from an overloaded server. A decentralized VM assignment and migration is presented in [38], which targets to make the resource management solution scalable. The decision regarding accepting new VMs is decided by servers (based on a probabilistic approach) inside datacenter based on their current utilization. Hierarchical resource management solution is another way of decreasing the complexity of the resource management solution. A hierarchical resource allocation solution to minimize the server energy consumption and maximize an SLA-based utility function for datacenters is presented in [31]. The proposed hierarchical solution breaks the problem of resource scheduling to multiple smaller problems (smaller server and application sets) to reduce the complexity of the problem and increase the parallelism.

Modeling the performance and energy cost is vital for solving the resource assignment problem. Good examples of theoretical performance modeling are [39,40]. Bennani and Menasce [39] present an analytical performance model based on queuing theory to calculate the response time of the clients based on CPU and I/O service times. Urgaonkar et al. [40]

present an analytical model for multitier internet applications based on the mean–value analysis. An example of experimental modeling of power and performance in servers is presented in [41].

In order to satisfy SLA and be able to keep the guaranteed level of performance for clients, datacenter resource manager needs to continuously monitor the performance of VMs, and utilization level of the active servers in order to perform VM migration to avoid possible SLA violation [42,43]. Different approaches are suggested in the literature to decide about the maximum utilization point at which VM migration needs to happen. For example, authors in [44,45] suggest to use the statistical CPU utilization behavior of the consolidated VMs on a server in order to come up with a workload behavior-adaptive utilization limit that triggers the VM migration to avoid SLA violation. This adaptive limit makes the decision regarding VM migration more accurate compared to a fixed maximum utilization limit.

Statistical analysis of the resource utilization is also used in order to decide about the VM consolidation in datacenter resource managers. For example, network bandwidth of VMs is dynamic and cannot be predicted perfectly. This fact motivated Wang et al. [46] to develop a solution to decide about VM consolidation based on the statistical data gathered from network bandwidth utilization of the VMs that can over-perform the VM consolidation solution based on the assumption of fixed communication bandwidth for each VM.

In order to be able to use the VM consolidation in its full extent, we need very fast VM migration solutions to avoid SLA violation in case of workload change. For example, Hirofuchi et al. [47] suggest using a fast solution called postcopy VM migration. In this approach, instead of migrating the whole memory before starting the VM operation in the destination host, VM operation starts right after the migration, and before memory copy is finished. In this case, if VM needs to access a point in its memory before all the memory is copied to the destination host, VM operation is stalled for a short amount of time before copy of that point of memory is finished. This solution significantly reduces the VM stall time during the live migration.

2.2 Power Management in Datacenters

Power management is one of the key challenges in datacenters. The power issue is one of the most important considerations for almost every decision-making process in a datacenter. In this context, the power issue refers to power distribution and delivery challenges in a datacenter, electrical energy cost due to average power consumption in the IT equipment and the room air conditioning, and power dissipation constraints due to thermal power budgets for VLSI chips.

Figure 2 depicts a distributed power management architecture composed of server-level power managers, plus blade enclosure and rack-level, and datacenter-level power provisioners, denoted as SPMs, EPPs, and DPP, respectively. There is one SPM per server, one EPP per blade enclosure, and a single DPP for the whole datacenter. This architecture is similar to the four-layer architecture proposed in [48]. The only difference with the architecture proposed in [48] is that instead of using one server power manager for each server that minimizes the average power consumption and avoids power budget violation, two power managers are proposed to do these jobs.

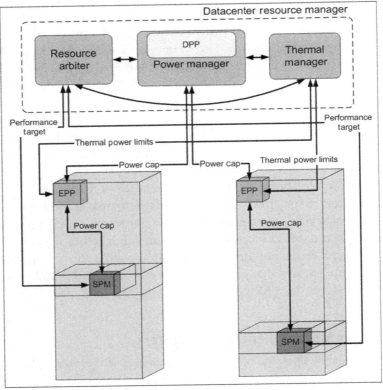

Figure 2 An example power management architecture and its relation to resource arbiter and thermal manager.

A number of dynamic power provisioning policies have been presented in the literature, including [48–50], where the authors propose using dynamic (as opposed to static) power provisioning to increase the performance in datacenter and decrease power consumption. Notice that the power provisioning problem can be formulated as deciding how many computing resources can be made active with a given total power budget for the datacenter.

Fan *et al.* [49] present the aggregate power usage characteristics of different units (servers, racks, clusters, and datacenter) in a datacenter for different applications over a long period of time. This data is analyzed in order to maximize the use of the deployed power capacity in the datacenter while reducing the risk of any power budget violations. In particular, this reference shows that there is a large difference between theoretical peak and actual peak power consumptions for different units. This difference grows as the unit size grows. This shows that the opportunity of minimizing the power budget under performance constraints (or maximizing the number of servers that are turned ON under a fixed power budget) increases as one goes higher in the datacenter hierarchy (e.g., from individual servers to datacenter as a whole). For example, it is reported that in a real Google datacenter, the ratio of the theoretical peak power consumption to actual maximum power consumption is 1.05, 1.28, and 1.39 for rack, power distribution unit (PDU), and cluster, respectively. The authors consider two approaches usually used for power and energy saving in datacenters, i.e., DVFS and reducing the idle power consumption in servers and enclosures (for example, by power gating logic and memory). Reported results suggest that employing the DVFS technique can result in 18% peak power reduction and 23% total energy reduction in a model datacenter. Moreover, decreasing the idle power consumption of the servers to 10% of their peak power can result in 30% peak power and 50% energy reduction. Based on these analyses and actual measurements, the authors present a dynamic power provisioning policy for datacenters to increase the possibility of better utilization of the available power while protecting the power distribution hierarchy against overdraws.

Exploring the best way of distributing a total power budget among different servers in a server farm in order to reach the highest performance level is studied in [51]. Moreover, an approach to reduce the peak power consumption of servers by dynamic power allocation using workload and performance feedbacks is presented in [52].

Design of an effective server-level power management is perhaps the most researched power management problem in the literature. Various dynamic power management (DPM) techniques that solve versions of this problem have been presented by researchers. These DPM approaches can be broadly classified into three categories: ad hoc [53], stochastic [54], and learning-based methods [55].

Server-level power manager can be quite effective in reducing the power consumption of datacenter. As an example, Elnozahy et al. [56] present independent as well as coordinated voltage and frequency scaling and turn ON/OFF policies for servers in a datacenter and compare them against each other from a power savings perspective. Their results indicate that independent DVFS policies for individual servers results in 29% power reduction compared to a baseline system with no DVFS. In contrast, a policy that considers only turning ON/OFF servers results in 42% lowering of the power consumption. The largest power saving of 60% is reported for a policy with coordinated DVFS and dynamic server ON/OFF decisions.

DPM techniques typically try to put the power consuming components to idle mode as often as possible to maximize the power saving. Studies on different datacenter workloads [7,49,57] show frequent short idle times in workload. Because of the short widths of these idle times, components cannot be switched to their deep sleep modes (which consume approximately zero power) considering the expected performance penalty of frequent go-to-sleep and wakeup commands. At the same time, because of energy nonproportionality of current servers [7], idle server power modes give rise to relatively high power consumption compared to the sleep mode power consumption. As discussed at length before, VM consolidation is an answer to this problem. A new solution is however emerging. More precisely, a number of new architectures have been presented for hardware with very low (approximately zero) idle mode power consumption (energy-proportional servers) to be able to reduce the average power consumption in case of short idle times [4,49].

There are many examples of work that describe a combined solution for power and resource management solution. For example, Wang and Wang [58] present a coordinated control solution that includes a cluster-level power control loop and a performance control loop for every VM. These control loops are configured to achieve desired power and performance objectives in the datacenter. Precisely, the cluster-level power controller monitors the power consumption of the servers and sets the DVFS state of the servers to reach the desired power consumption. In the same venue, the VM performance controller dynamically manages the VM performance by changing the resource (CPU) allocation policy. Finally, a cluster-level resource coordinator is introduced whose job is to migrate the VMs in case of performance violation. As another example, Buyya and Beloglazov [59] propose a management architecture comprising of a VM dispatcher, as well as local and global managers. A local manager migrates a VM from one server to another in case of SLA violations, low server utilization, high server temperature, or high amount of communication with another VM in a different server. A global manager receives information from local managers and issues commands for turning ON/OFF servers, applying DVFS, or resizing VMs.

This chapter tackles the resource management problem in a cloud computing system. Key features of our formulation and proposed solution are that we consider heterogeneous servers in the system and use a two-dimensional model of the resource usage accounting for both computational and memory bandwidth. We propose multiple copies of VMs to be active in each time in order to reduce the resource requirement for each copy of the VM and hence help to increase the chances for VM consolidation. Finally, an algorithm based on DP and local search is described. This algorithm determines the number of copies of each VM and the placement of these copies on servers so as to minimize some total system cost function.

3. SYSTEM MODEL

In this section, detail of the assumptions and system configuration for the VM placement problem are presented. To improve the readability, Table 1 presents key symbols and definitions used in this chapter. Note that each client is identified by a unique id, denoted by index i whereas each server in the cloud computing system is identified by a unique id, denoted by index j.

3.1 Cloud Computing System

In the following paragraphs, we describe the type of the datacenter that we have assumed as well as our observations and key assumptions about where the performance bottlenecks are in the system and how we can account for the energy cost associated with a client's VM running in the datacenter.

A datacenter comprises of a number of potentially heterogeneous servers chosen from a set of known and well-characterized server types. In particular, servers of a given type are modeled by their processing capacity or CPU cycles ($C_\star^p$) and memory bandwidth ($C_\star^m$) as well as their operational expense (energy cost), which is directly related to their average power consumption.

Table 1 Notation and Definitions

Symbol Name	Definition
c_i^m and c_i^p	Required memory bandwidth and total processing demand of the ith client
L_i	Max. number of servers allowed to serve the ith client
s_k	Set of servers of type k
C_j^p and C_j^m	Total CPU cycle capacity and memory bandwith of the jth server
P_j^0	Constant power consumption of the jth server in the active mode
P_j^p	Power of operating the jth server which is proportional to the utilization of processing resources
T_e	Duration of a epoch in seconds
x_j	A pseudo–Boolean variable to determine if the jth server is ON (1) or OFF (0)
y_{ij}	A pseudo–Boolean variable to determine if the ith VM is assigned to the jth server (1) or not (0)
ϕ_{ij}^p, ϕ_{ij}^m	Portion of the processing and memory bandwidth resources of the jth server that is allocated to the ith client
ϕ_j^p, ϕ_j^m	Portion of the processing and memory bandwidth resources of the jth server that is allocated to any clients
α	Processing size ratio of the VM copy (between $1/L_i$ and 1) that determines the portion of the original VM CPU cycle provided by the VM copy
$f(\alpha)$	Function of processing size ratio that is used in calculating ϕ_{ij}^p based on c_i^p amd C_j^p
$c_{ij}(\alpha)$	Estimate of the energy cost of assigning a copy of the ith VM with processing size ratio of α to the jth server
y_{ij}^α	assignment parameter for jth server with VM with processing size ratio of α

We assume that local (or networked) secondary storage (disk) is not a system bottleneck.

The main part of the operational cost of the system is the total energy cost of serving clients' requests. The energy cost is calculated as the server power multiplied by the duration of each epoch in seconds (T_e). The power dissipation of a server is modeled as a constant power cost $(P_\star^0)$ plus another variable power cost, which is linearly related to the utilization of the server (with slope of $P_\star^p$). This model is inspired by the previous works such as [17,41]. Note that the power cost of communication resources and air conditioning units are amortized over all servers and communication/networking gear in the datacenter, and are thus assumed to be relatively independent of the clients' workloads. More precisely, these costs are not included in the equation for power cost of the datacenter.

3.2 Client and VMs

Clients in the cloud computing system are represented as VMs. Based on the SLA contract or using workload prediction with consideration of the SLA, the amount of resources required for each client can be determined. These VMs are thus considered to have processing and memory bandwidth requests during the considered epoch. This assumption is applicable to online services (not for batch applications).

Each client's VM may be copied on different servers (i.e., requests generated by a single VM can be assigned to more than one server). This request distribution can decrease the quality of the service if the number of servers that processes the client requests is large [30]. Therefore, we impose an upper bound on this number; precisely, L_i determines the maximum number of copies of any VM in the datacenter (this bound can be set to one if for some reason the VM should not be replicated). When multiple copies of a VM are active on different servers, the following constraints must be satisfied:

$$\sum_j \phi_{ij}^p C_j^p \geq c_i^p \tag{1}$$

$$\phi_{ij}^m C_j^m = \gamma_{ij} c_i^m \tag{2}$$

where ϕ_{ij}^p and ϕ_{ij}^m denote the portion of the jth server CPU cycles and memory bandwidth allocated to the VM associated with client i. Constraint (1) enforces the summation of the reserved CPU cycles on the assigned servers to be equal or greater than the required CPU cycles for client i. Constraint (2) enforces the provided memory bandwidth on assigned servers to be equal to the required memory bandwidth for the VM. This constraint enforces the cloud provider not to sacrifice the QoS of clients. An example of VM1 being replicated as VM2 and VM3 is shown in Fig. 3.

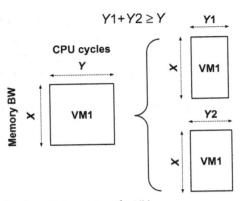

Figure 3 An example of multiple copies of a VM.

3.3 VM Management System

The focus of the rest of this chapter is on VM manager, which is responsible for determining resource requirements of the VMs and placing them on servers. Moreover, to address dynamic workload changes, VM manager may do VM migration. VM manager performs these tasks utilizing two different optimization procedures: semistatic optimization and dynamic optimization. The semistatic optimization procedure is performed periodically, whereas the dynamic optimization procedure is performed whenever it is needed.

In the semistatic optimization procedure, VM manager considers the full active set of VMs, the previous assignment solution, feedbacks generated by the power, thermal, and performance sensors, and workload prediction in order to generate the best VM placement solution for the next epoch. The period for performing semistatic optimization depends on the type and size of the datacenter and workload characteristics. In the dynamic optimization procedure, VM manager finds a temporary VM placement solution by migrating, creating, or removing some VMs in response to any performance, power budget, or critical temperature violations.

In this work, we focus on semistatic optimization procedure of VM manager. In this procedure, resource requirements of VMs are assumed to be determined based on SLA contracts and workload estimation for the next epoch. The duration of the epoch is long enough for one to neglect the VM migration delay penalty (it is typically less than 100 ms for live migration [23]) with respect to the gain of the global optimization. Consequently, the energy cost optimization may be performed without the constraint of the state of the cloud computing system in the previous epoch.

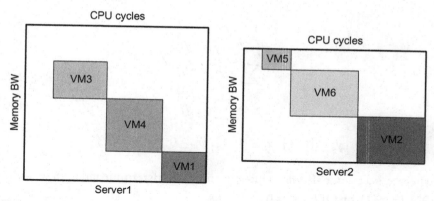

Figure 4 An exemplary solution for assigning six VMs on two different servers.

The role of semistatic optimization procedure in VM manager is to answer the questions of (i) whether to create multiple copies of VMs on different servers or not and (ii) where to place these VMs. Considering fixed payments from clients for the cloud service, the goal of this optimization is to minimize the operational cost of the active servers in datacenter. An exemplary solution for assigning six VMs on two heterogeneous servers is shown in Fig. 4.

4. PROBLEM FORMULATION

In this chapter, a VM placement problem is considered with the objective of minimizing the total energy consumption in the next epoch while servicing all VMs in the cloud computing system.

The exact formulation of the aforesaid problem (called EMRA for energy-efficient multi-dimensional resource allocation) is provided below (cf. Table 1)

$$\text{Min } T_e \sum_j x_j \left(P_j^0 + P_j^p \sum_i \phi_{ij}^p \right) \tag{3}$$

subject to:

$$\phi_j^p = \sum_i \phi_{ij}^p \leq 1 \ \forall j \tag{4}$$

$$\phi_j^m = \sum_i \phi_{ij}^m \leq 1 \ \forall j \tag{5}$$

$$\sum_j C_j^p \phi_{ij}^p \geq c_i^p \ \forall i, j \tag{6}$$

$$y_{ij} \geq \phi_{ij}^p, \ \forall i, j \tag{7}$$

$$\phi_{ij}^m C_j^m = y_{ij} c_i^m \ \forall i, j \tag{8}$$

$$\sum_i y_{ij} \leq L_i \ \forall i \tag{9}$$

$$x_j \geq \sum_i \phi_{ij}^p \ \forall j \tag{10}$$

$$y_{ij} \in \{0, 1\}, x_j \in \{0, 1\}, \phi_{ij}^p \geq 0, \phi_{ij}^m \geq 0 \ \forall i, j \tag{11}$$

where x_j is a pseudo-Boolean integer variable to determine if the jth server is ON $(x_j = 1)$ or OFF $(x_j = 0)$.

The objective function is the summation of the operation costs (energy dissipations) of the ON servers based on a fixed power factor and a variable power term linearly related to the server utilization. In this problem, x_j, y_{ij}, and ϕ_{ij}^p denote the optimization variables.

The constraints capture the limits on the number of available servers and clients. In particular, inequality constraints (4) and (5) represent the limit on the utilization of the processing and memory bandwidth in the jth server, respectively. Constraint (6) ensures that required processing demands for each VM is provided. Constraint (7) generates a pseudo-Boolean parameter that determines if a copy of a VM is assigned to a server $(y_{ij} = 1)$ or not $(y_{ij} = 0)$. Constraint (8) ensures the memory bandwidth needs of a VM that is assigned to a server are met whereas constraint (9) ensures that the number of copies of a VM does not exceed the maximum possible number of copies. Constraint (10) generates the pseudo-Boolean parameter related to the status of each server. Constraint (11) specifies the domains of optimization variables.

Theorem I. Generalized assignment problem (GAP) [24] can be reduced to EMRA problem.

Proof. Consider a version of the EMRA problem in which P_j^0 is equal to zero for every server and L_i is equal to 1 for every VM. In this problem, assigning each VM (exactly one copy) to each server has different costs, and each server has two-dimensional resources that can be assigned to VMs. So, we can solve any two-dimensional GAP problem by using the solution to EMRA problem for a special case.■

Considering Theorem I, the EMRA problem is NP-hard [24]. Indeed, similar to the GAP problem, even the question of deciding whether a feasible solution exists for this problem does not admit an efficient solution [24].

In this chapter, we consider a case in which the required resources for VMs are smaller than the available resources in the datacenter. This means that we consider energy minimization with a fixed set of VMs instead of maximizing the number of (or the profit for) a subset of VMs served in the datacenter. Therefore, a simple greedy algorithm (similar to FFD heuristic [24]) will find a feasible solution to the EMRA problem. Another important observation about this problem is that the number of clients and servers in this problem are very large; therefore, a critical property of any proposed heuristic is its scalability.

An example of how multiple copies of VM can reduce energy consumption of the cloud system is seen when we compare Fig. 5A and B. Here, three homogenous VMs are assigned to three homogenous servers. The CPU cycle capacity of each server is strictly less than 2 × the required CPU cycle count of each VM (say, 1.75 ×) whereas the memory bandwidth capacity of each server is strictly more than 2 × the required memory bandwidth of each VM (say, 3 ×). In Fig. 5, you can see that the assignment results in three active servers. If we consider VM replication, we can create two copies of the third VM with the same memory bandwidth requirements but smaller

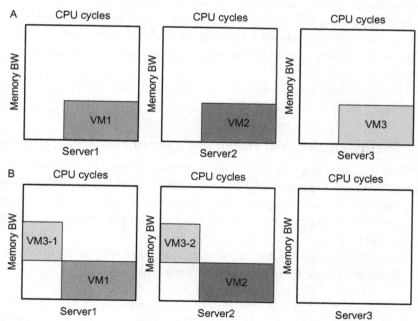

Figure 5 An exemplary solution for assigning three VMs on three identical servers. (A) Without VM replication and (B) with VM replication.

CPU cycle demands. Assigning the new set of VMs to servers can result in only two active servers with high CPU and memory bandwidth utilization, which may result in energy saving due to the energy nonproportionality behavior of the servers.

5. PROPOSED ALGORITHM

In this section, a two-step heuristic for solving the EMRA problem is presented. In the first step, an algorithm based on DP is used to determine the number of copies for each VM and the assignment of these VMs to the servers. This decision determines (i) which servers will be active during the next epoch and (ii) the utilization of the active servers in that epoch. The goal of the algorithm is to minimize the total energy cost. In the second step, a local search is conducted to further reduce the power consumption by turning off some of the active servers and placing their VMs on other active servers.

In the beginning of the VM placement, clients are ordered in descending order of their CPU cycle demands. Based on this ordering, the optimal number of copies of the VMs are determined and they are placed on servers by using DP. In the local search method, servers are turned off based on their utilization and VMs assigned to them are moved to the rest of the active servers so as to minimize the energy consumption as much as possible.

Details of the Energy-efficient VM Placement algorithm (EVMP) are presented below.

5.1 Energy-Efficient VM Placement Algorithm—Initial Solution

Initially, the values of ϕ_j^p and ϕ_j^m for each server are set to zero. A constructive approach is used to place the VMs on the servers. VMs are sorted based on their processing requirements in a descending order. For each VM, a method based on DP is used to determine the number of copies that are placed on different servers.

To estimate the power consumption of assigning a copy of the ith VM to the jth server of type k, we use the following equation

$$c_{ij}(\alpha) = \begin{cases} T_e\left(\phi_{ij}^p P_j^p + P_j^0 c_i^m / C_j^m\right) & \text{If server is active} \\ T_e\left(\phi_{ij}^p P_j^p + P_j^0\right) & \text{otherwise} \end{cases} \tag{12}$$

where α (between $1/L_i$ and 1) is the processing size ratio of the VM copy to that of the original VM. In other words, α denotes the percentage of the original VM CPU cycles to be provided to the copy of VM.

The top branch of Eq. (12) estimates the energy cost of assigning a copy of the ith VM to an already active server and is comprised of a utilization-proportional power consumption of the server plus a fraction of the idle power consumption based on the (normalized) required memory bandwidth of the assigned VM. Similarly, the bottom branch of Eq. (12) estimates the energy cost of assigning a copy of the ith VM to a currently inactive server (but to become active soon). This energy cost estimate includes the utilization-proportional power consumption of the server and accounts for the whole idle power consumption of the server. The additional power consumption in the bottom branch compared to the one in the top branch captures the risk of turning on a server if no other VM is assigned to that server for the next epoch. The average utilization of the server type ($\overline{\phi}_j^p$ and $\overline{\phi}_j^m$) in the previous epochs can also be used to replace the equation in the bottom branch with $T_e\left(\phi_{ij}^p P_j^p + P_j^0 \min\left(1, \max\left(c_i^m/\overline{\phi}_j^m, c_i^m/\overline{\phi}_j^m\right)\right)\right)$ to more accurately account for the energy cost risk of turning on a server for VM assignment.

ϕ_{ij}^p is a function of the VM, the server, and α. It can be calculated as shown below.

$$\phi_{ij}^p = f(\alpha)c_i^p / C_j^p \qquad (13)$$

where $f(\alpha)$ is a function of the processing size ratio of the VM. We know that in any type of VM and servers, $f(0)$ is equal to 0 while $f(1)$ is equal to 1. f is a monotonically increasing function. Considering the beginning and endpoint of this function at 0 and 1 and considering constraint (1), for any value between 0 and 1, the value of function f can be between α and 1. For example, if half of the CPU cycle requirement of the VM is provided by a copy of the VM, $\phi_{ij}^p = f(1/2)c_i^p / C_j^p$ which is greater than or equal to $0.5c_i^p / C_j^p$. If this property does not hold for a small portion of the spectrum, we can create a solution with multiple VM copies which require less than c_i^p resources collectively and violate constraint (1).

The presented algorithm is based on a general function f with the mentioned behavior but an example of this function based on a performance model is presented in Section 5.2.

For each VM, both versions of Eq. (12) are calculated for each server type and different values of α (between $1/L_i$ and 1 with steps of $1/L_i$). Moreover, for each server type, L_i active servers and L_i inactive servers that can service at least the smallest copy of the VM are selected as candidate hosts. For assigning the VM to any of the candidate servers, the cost is determined by the top or bottom branch of Eq. (12) as the case may be.

After selecting active and inactive candidate servers for each server type and calculating cost for each possible assignment, the problem is reduced to (14).

$$\text{Min} \sum_{j \in P} y_{ij}^\alpha c_{ij}(\alpha) \tag{14}$$

subject to:

$$\sum_{j \in P} \alpha y_{ij}^\alpha = L_i \tag{15}$$

where y_{ij}^α denotes the assignment parameter for jth server for a VM copy with processing size ratio of α (1 if assigned and 0 otherwise). Moreover, P denotes the set of candidate servers for this assignment.

The DP method is used to solve this problem and find the best assignment decision. In this DP method, candidate servers can be processed in any order. This method examines all the possible VM placement solution efficiently without calculating every possible solution in a brute-force manner. Using this method, the optimal solution for problem presented in (14) can be found.

Algorithm 1 shows the pseudo code for this assignment solution for each VM. Complexity of this DP solution is $O(2L_i^2 K)$, where K denotes the number of server types that are considered for this assignment. The complexity is calculated from the number of cost calculation in line 23 of the pseudo code. After finding the assignment solution ϕ_j^p and ϕ_j^m for the selected servers are updated. Then, the next VM is chosen and this procedure is repeated until all VMs are placed.

Algorithm 1. *Energy-Efficient VM Placement*
Inputs: C_j^m, C_j^p, P_j^0, P_j^p, c_i^m, c_i^p, L_i
    Outputs: ϕ_{ij}^p, ϕ_{ij}^m (i is constant in this algorithm)
    1   $P = \{\}$
    2   **For** ($k = 1$ to number of server types)
    3      $ON = 0$; $OFF = 0$;
    4   **For** ($\alpha = 1/L_i$ to L_i)
    5         $\phi_{ij}^p = f(\alpha) c_i^p / C_j^p$
    6         $c_{ij}^{active}(\alpha) = \phi_{ij}^p P_j^p + P_j^0 c_i^m / C_j^m$

7      $c_{ij}^{inactive}(\alpha) = \phi_{ij}^p P_j^p + P_j^0$

8   **End**

9      $J^{ON} = \left\{ j \in s_k \middle| \left(1 - \phi_j^m\right) \geq c_i^m / C_j^m \,\&\, \left(1 - \phi_j^p\right) \geq c_i^p / L_i C_j^p \right\}$

10      $J^{OFF} = \left\{ j \in s_k \middle| \phi_j^p = 0, \left(1 - \phi_j^m\right) \geq c_i^m / C_j^m \right\}$

11   **Foreach** $(j \in s_k)$

12   **If** $(j \in J^{ON} \,\&\, ON < L_i)$

13          $P = P \cup \{j\},\ ON{+}{+}, c_{ij}(\alpha) = c_{ij}^{active}(\alpha)$

14   **Else if** $(j \in J^{OFF} \,\&\, OFF < L_i)$

15          $P = P \cup \{j\},\ OFF{+}{+},\ c_{ij}(\alpha) = c_{ij}^{inactive}(\alpha)$

16   **End**

17   **End**

18      $X = L_i$, and $Y = size(P)$

19   **Foreach** $(j \in P)$

20   **For** $(x = 1 \text{ to } X)$

21        $D[x, y] = \text{infinity};$ // Auxiliary $X \times Y$ matrix used for DP

22   **For** $(z = 1 \text{ to } x)$

23          $D[x, y] = min\left(D[x, y], D[x-1, y-z] + c_{ij}(z)\right)$

24          $D[x, y] = min\left(D[x, y], D[x-1, y]\right)$

25   **End**

26   **End**

27   Back-track to find best ϕ_{ij}'s to minimize cost and update ϕ_j's

5.2 Example of Function $f(\alpha)$ for a Performance Model

To better appreciate the concept of function $f(\alpha)$, a performance model for VM is briefly presented.

To model the response time of the VMs, we assume that the interarrival times of the requests for each VM follow an exponential distribution function similar to the interarrival times of the requests in the e-commerce applications [28]. The average interarrival time (λ_i) of the requests for each VM can be estimated from analyzing workload traces [60].

In case of more than one copy of a VM, requests are assigned probabilistically, i.e., α portion of the incoming requests are forwarded to the jth server (i.e., the host for some copy of the VM) for execution, independently of the past or future forwarding decisions. Based on this assumption, the request arrival rate in each server follows the Poisson distribution function.

An exponential distribution function can be used to model the service time of the clients in this system. Based on this model, the response time distribution of a VM (placed on server j) is an exponential distribution with the following expected value:

$$\bar{R}_{ij} = \frac{1}{C_j^p \phi_{ij}^p \mu_{ij} - \alpha \lambda_i} \tag{16}$$

where μ_{ij} denotes the service rate of the ith client on the jth server when a unit of processing capacity is allocated to the VM of this client.

Most response time sensitive applications have a contract with the cloud provider to guarantee that the response time of their requests does not go over a certain threshold. The constraint on the response time of the ith client may be expressed as:

$$\text{Prob}\{R_i > R_i^c\} \le h_i^c \tag{17}$$

where R_i and R_i^c denote the actual and target response times for the ith client's requests, respectively.

Based on the presented model and constraint (17), the response time constraint for each copy of a VM can be expressed as follows:

$$e^{-\left(C_j^p \phi_{ij}^p \mu_{ij} - \alpha \lambda_i\right) R_i^c} \le h_i^c \Rightarrow \phi_{ij}^p \ge \left(\alpha \lambda_i - \ln h_i^c / R_i^c\right) / \mu_{ij} C_j^p \tag{18}$$

If there is only one copy of VM, c_i^p can be calculated as follows:

$$c_i^p = \lambda_i / \mu_{ij} C_j^p + \left(-\ln h_i^c / R_i^c\right) / \mu_{ij} C_j^p \tag{19}$$

Considering the presented performance model, c_i^p varies based on the server type. If the processing size ratio of α is considered for the VM copy, lower bound of ϕ_{ij}^p has a similar formula as (19) with the first term multiplied by α. The first term of ϕ_{ij}^p is the portion that scales with the processing size ratio of the VM. The second term of ϕ_{ij}^p is a constant value based on the SLA contract parameters, service rate, and processing capacity of the server. Note that the second term does not scale with α and exists in even the smallest VM copy to guarantee that the request is serviced with an acceptable response time.

Having multiple copies of VM requires to account for the second term multiple times. For example, if there are three active copies of a VM, independent of the value for α parameter for each copy, the summation of ϕ_{ij}^p

is equal to $\lambda_i/\mu_{ij}C_j^p + 3 \times \left(-\ln h_i^c/R_i^c\right)/\mu_{ij}C_j^p$. This value would be larger than the c_i^p. The function f for this performance model is presented below.

$$f(\alpha) = \frac{\alpha\lambda_i/\mu_{ij}C_j^p + \left(-\ln h_i^c/R_i^c\right)/\mu_{ij}C_j^p}{\lambda_i/\mu_{ij}C_j^p + \left(-\ln h_i^c/R_i^c\right)/\mu_{ij}C_j^p}$$

$$= \alpha + \frac{(1-\alpha)\left(-\ln h_i^c/R_i^c\right)/\mu_{ij}C_j^p}{\lambda_i/\mu_{ij}C_j^p + \left(-\ln h_i^c/R_i^c\right)/\mu_{ij}C_j^p} \tag{20}$$

The behavior of f is determined from the ratio between the first and second terms in Eq. (19). When this ratio is big, creating a limited number of copies from that VM is reasonable since $f(\alpha) \cong \alpha$ and the total amount of processing power reserved and used for multiple copies of VM is approximately equal to the processing power needed for the original VM. On the other hand, when the aforesaid ratio is small, then the VM is not a good candidate for replication since the total processing power required for multiple copies of that VM is multiple times larger than the required processing power for the original VM. However, as shown before, in some scenarios the increase in utilization of servers and turning off some other servers by creating multiple copies of VM can decrease the overall operational cost of the datacenter and cloud system. The proposed algorithm can capture this trade-off and come up with the near optimal solution.

5.3 Energy-Efficient VM Placement Algorithm—Local Search

The constructive nature of the proposed algorithm can cause a situation in which some servers are not well utilized. However, the large number of clients makes this problem less severe. To improve the results of the proposed VM placement algorithm, a local search method is used.

In order to select the candidate servers for turning OFF, utilization of the server needs to be defined. Due to heterogeneity of the server resources and VM resource requirements, it is possible that the utilization ratio of the server along different resource dimensions will be different. Since saturation of each resource type in the server results in a resource-saturated server (to be called a fully utilized server), we define the utilization of a server as the maximum resource utilization along different resource dimensions. For example if $\phi_j^p = 0.5$ and $\phi_j^m = 0.3$, we consider the utilization of the server to be 50%. To minimize the total energy consumption in the system, all servers with utilization less than a threshold will be examined in this local search. This threshold can be specified by the cloud provider.

To examine these under-utilized servers, each of them is turned off one by one (starting from servers with the lowest utilization) and total energy consumption is found by placing their VMs on other active servers using the proposed DP placement method. If the total cost of the new placement is less than the previous total cost, the new configuration is fixed, and the rest of under-utilized servers are examined; otherwise, the option of turning off that server is removed and the other candidate servers are examined. Algorithm 2 shows a high-level pseudo code for the proposed local search step.

Algorithm 2. *Local Search Algorithm*

Inputs: Current VM assignment and x_j

   Outputs: New VM assignment and x_j

1   $\phi_j = max\left(\phi_j^m, \phi_j^p\right)$

2   $J = \left\{j | \phi_j > 0\right\}$

3   **While** ($\phi_j <$ threshold OR timeout)

4      $j = \text{argmin}_{j \in J | \phi_j > 0} \phi_j$

5      $I = \left\{i | \phi_{ij}^m > 0\right\}$

6      $OPEX_{old} =$ Total operational cost based on the current assignment

7   **Foreach** ($i \in I$)

8   Find a new placement on set of active servers

9   **End**

10    $OPEX_{new} =$ Total operational cost based on the new assignment

11   If ($OPEX_{new} < OPEX_{old}$)

12        $x_j = 0$ and fix the new VM assignment

13   Else

14        $x_j = 1$ and keep the old VM assignment

15    $J = J - j$

16 **End**

17 Finalize the set of active servers and VM assignment for the current epoch

6. SIMULATION RESULTS

To evaluate the effectiveness of the proposed VM placement algorithm, a simulation framework is implemented. Simulation setups, baseline heuristics, and numerical results of this implementation are presented in this section.

6.1 Simulation Setup

For simulation purposes, model parameters are generated from real world examples. The number of server types is set to 8. For each server type, some arbitrary number of servers are provisioned in datacenter. Processors for each server type are selected from the Intel portfolio of processors (e.g., Atom, i5, i7, and Xeon) [61] with different number of cores, cache sizes, power consumptions, and clock frequencies. Peak power consumptions for different servers (excluding the processor itself) are set uniformly to be between 2 and $4 \times$ the power consumption of the corresponding fully utilized processor. The memory bandwidth requirements of the servers are selected based on the maximum memory bandwidth of these processors multiplied by a factor of 0.4. For example, if the maximum memory bandwidth of a processor is 20 GB/s, the available memory bandwidth for this processor is set to 8 GB/s.

The processing (CPU cycle count) requirement for each VM is selected uniformly between 1 and 18 billion CPU cycles per second. In order to observe the effect of function $f(\alpha)$, we ran the experimental results twice for each setting. The first time considering $f(\alpha) = \alpha$ and the second time with $f(\alpha) = (\alpha + 1)/2$. f_1 and f_2 denote the first and second observed values of $f(\alpha)$. As described in Section 5.2, $f(\alpha)$ is a function of the type of VM and is not constant for all VMs in datacenter. The purpose of considering two different $f(\alpha)$ for the simulation setup is to show how the algorithm works with different VM replication costs.

The memory bandwidth requirements for clients are also selected uniformly between 768 MB/s and 4 GB/s. The selection of processing resource requirement is based on the fact that the baseline algorithms do not automatically support multiple copies of VMs. This means that the required processing capacity of each VM should be less than the maximum available processing capacity in the datacenter; otherwise, the baseline algorithms cannot handle the VM placement problem. On the other hand, EVMP algorithm is capable of generating a VM placement solution if the memory bandwidth requirement of each VM is less than the maximum memory bandwidth supported by the available servers in the datacenter.

Upper bound on the number of copies for each VM is set between 1 and 5 based on the value of the required processing resources, e.g., if the processing requirement for a VM is equal to maximum processing requirements, L_i is set to 5 and if the value of processing requirement for a VM is less than ¼ of the maximum value, L_i is set to one (no copy is allowed).

Each simulation is repeated at least $1000 \times$ to generate acceptable average results for each case.

6.2 Heuristics for Comparison

We implemented the *min power parity* (mPP) heuristic [23] as one of the state-of-the-art energy-aware VM placement techniques. This heuristic is based on FFD heuristic [24] for the bin-packing problem. This heuristic tries to minimize the overall power consumed by active servers in the datacenter. mPP heuristic works in two steps. In the first step, a target utilization for each server is found based on the power model for the servers. The target utilization of the servers is found by minimizing the power consumption of assigning the total required CPU utilization of all VMs on the current servers. In the second step, FFD heuristic is used to assign VMs to the selected set of the active servers. More details of mPP can be found in [23].

To show the effectiveness of our proposed approach for placing multiple copies of VMs on servers, along with mPP, a version of our algorithm in which L_i is set to one for all i is also considered. We refer to this version of the algorithm with the name of *baseline method* in the figures.

Moreover, to show the effect of distributed resource assignment and constant power cost for active servers, we implement a procedure to find a lower bound on the total energy cost with relaxation of these obstacles. To calculate this lower bound, for each VM, total energy cost $\left(c_i^p / C_j^p \left(P_j^p + P_j^0 \right) T_e \right)$ of serving that VM on each server is calculated and the smallest energy cost is selected. Summation of these energy costs generates a lower bound on the total energy cost in the system.

6.3 Numerical Results

Normalized total energy cost in the system using the EVMP algorithm, baseline method, and mPP algorithm is presented in Fig. 6. EVMP-f1 results show the results for the first $f(\alpha)$ function whereas EVMP-f2 shows the results for the second $f(\alpha)$ function as discussed earlier.

As can be seen, EVMP reduces the total energy cost of VM placement solution by 24–36% with respect to the mPP algorithm. This amount of energy decrease is significant in cloud computing systems and can help reduce the operational cost of computing.

As can be seen, changing $f(\alpha)$ function from $f1$, which represent ideal VM copying case, to $f2$, which captures a scenario in which the cost of creating multiple copies of VMs is rather large, does not significantly increase the

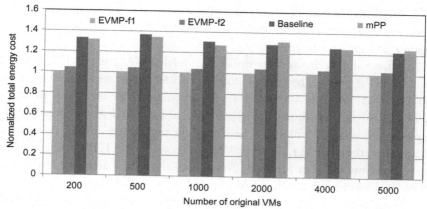

Figure 6 Normalized total energy cost of the system.

Table 2 Performance of the EVMP-F1 w.r.t. Lower Bound Cost and Average Number of VM Copies

Number of Original VMs	Performance w.r.t. Lower Bound
200	1.02
500	1.01
1000	1.05
2000	1.01
4000	1.08
5000	1.05

energy consumption of the datacenter (between 3% and 4% increase). This is due to the fact that EVMP algorithm adapts the decision regarding VM copying based on the $f(\alpha)$ function. This means that having a higher cost associated with creating copies of some VM, results in fewer number of VM copies being created by the algorithm.

Performance of the baseline algorithm which is based on assigning VMs using DP method is slightly worse than the performance of mPP method (~3% range) because baseline method does not place the VM on the server with the least resource availability and instead chooses the host server randomly in the selected server type.

Table 2 shows the relative performance of EVMP-f1 with respect to the derived lower bound on the total energy cost. There are two reasons behind the difference between the result of EVMP and the lower bound:

(i) imperfection of the algorithm and (ii) constant power consumption of the servers (independent from their utilization) and effect of the distributed resources in the datacenter.

Table 3 shows the average number of VM copies created in EVMP-f1 and EVMP-f2 runs. The average number of VM copies on the final solution of the EVMP-f1 and EVMP-f2 is small compared to the average L_i for VMs which is three. This shows that the EVMP algorithm does not create multiple copies of a VM unless it is beneficial for the energy cost of the system. Moreover, the average number of VM copies created in EVMP-f2 is smaller than the same number for EVMP-f1 which shows the adaptiveness of the EVMP algorithm in creating VM copies based on $f(\alpha)$ function.

Effect of different L_i values on the performance of EVMP-f1 is reported in Fig. 7. In this figure, the normalized total energy costs of the VM placement solutions when using the EVMP algorithm and for different L_i values are shown. As can be seen, the cost difference between the EVMP solution and the solution of a version of EVMP that restricts the number of VM copies to two is 7% (on average). This shows around 20% energy reduction compared to the mPP algorithm even if the number of allowed VM copies is limited to two. The cost difference between the EVMP solution and the solution of a version of EVMP that restricts the number of VM copies to 10 is around 10% (on average). Note that the function used to calculate the resource requirement for each VM copies for EVMP-f1 only accounts for the lower bound amount of the processing resources required for each VM copy. Figure 8 shows the same comparison for EVMP using the second $f(\alpha)$ function. As can be seen, the energy cost reduction when the maximum allowable number of VM copies is increased to 10, is smaller (about 6% improvement) in case of using the second $f(\alpha)$ function. This is due to

Table 3 Performance of the EVMP-F2 w.r.t. Lower Bound Cost and Average Number of VM Copies

Number of Original VMs	Average No. of VM Copies for EVMP-F1	Average No. of VM Copies for EVMP-F2
200	1.83	1.76
500	1.78	1.71
1000	1.78	1.72
2000	1.74	1.67
4000	1.71	1.66
5000	1.69	1.64

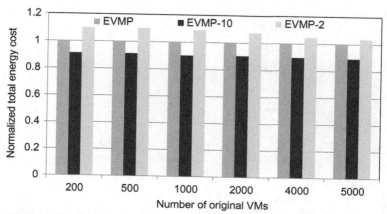

Figure 7 Normalized total energy cost of the VM placement solution using for different L_i for EVMP-f1.

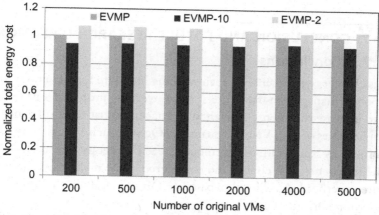

Figure 8 Normalized total energy cost of the VM placement solution using for different L_i for EVMP-f2.

the fact that f2 function adds an energy cost penalty every time a new VM copy is added to the system.

Figure 9 shows the average run-time of the EVMP, baseline, and mPP methods for different number of VMs. Note that VM placement algorithm is called only a few times in each charge cycle (1 h in Amazon EC2 service [62]), e.g., 2–3× per hour. Also to reduce the time complexity of the EVMP algorithm in case of bigger number of VMs, we can use a partitioning algorithm to assign a set of VMs to a cluster and then apply EVMP in each cluster in parallel.

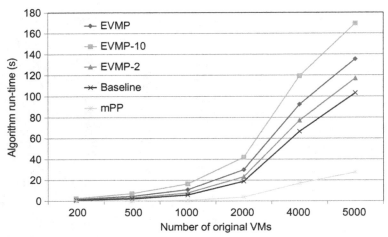

Figure 9 Run-time of EVMP for different number of VMs on 2.4 GHz E6600 server with 3 GB of RAM from Intel.

7. CONCLUSION AND FUTURE RESEARCH DIRECTION

7.1 Conclusions

In this chapter, we presented a review of the literature focusing on resource and power managers in datacenters. Moreover, we proposed a novel solution to increase the energy efficiency in datacenter that relies on generating multiple copies of each VM. To guarantee QoS for each VM, we considered fixed memory bandwidth requirement for each VM copy, added a limitation on the number of VM copies, and considered a VM replication energy and resource overhead. An algorithm based on DP and local search was proposed to determine the number of VM copies, then place them on servers to minimize the total energy cost in the cloud computing system. Using simulation results, we showed that this approach reduces the total energy cost $\sim 20\%$ with respect to the prior VM placement techniques. The effect of different parameters on the system performance was also evaluated using simulation results.

The proposed solution provides a flexible method to increase the energy efficiency of the cloud computing system and increases the resource availability in the datacenter. Cloud provider can decide how to service VMs with big processing resource requirements and how to distribute their requests among the servers to maximize the energy efficiency.

7.2 Possible Research Direction on Energy-Efficient Datacenter Design

There are plenty of opportunities to improve the state of the art in resource and power managers in datacenters. Advancing the design and adaptive control of datacenters with energy efficiency, SLAs, and total cost of ownership are the primary areas that one can contribute on, as detailed below.

The first step is to develop a theory for understanding the energy complexity of computational jobs. Today, energy efficiency is benchmarked relative to last year's product; any efficiency gain is touted as success. Instead, we wish to ask what level of efficiency is possible and measure solutions relative to this limit. One must thus develop key scientific principles to measure the energy complexity of applications. By combining energy complexity with time complexity of applications, we can then perform fundamental energy-performance trade-offs at application programming level.

Informed by this new theory, one can then reconsider the design of the hardware platforms that comprise the energy-efficient datacenters. Key sources of inefficiency are the lack of energy-proportional hardware and the overprovisioning of these servers to meet SLAs given the time-varying application resource demands. An energy-efficient datacenter exploits hardware heterogeneity and employs dynamic adaptation. Heterogeneity allows energy-optimized components to be brought to bear as an application characteristics change. Dynamic adaptation allows the datacenter to adapt and provision hardware components to meet varying workload and performance requirements, which, in turn, eliminates overprovisioning. Computing, storage, and networking subsystems of current datacenters exhibit dismal energy proportionality. One must attempt to redesign server architectures and network protocols with energy efficiency and energy proportionality as the driving design constraint. On the storage front, we must construct hybrid storage systems that assign data to devices based on a fundamental understanding of access patterns and capacity–performance–efficiency trade-offs.

To go beyond the incremental energy efficiency gains possible from component-wise optimization, one must consider the coordination and control of storage, networking, memory, compute, and physical infrastructure. By tackling the optimization problem for the datacenter as a whole, one can develop solutions at one layer that will be exploited at other layers. By using the mathematical underpinnings of control theory and stochastic modeling, these approaches enable reasoning about worst-case and average-case behavior of multiloop compositions of control approaches. One can then develop algorithms to globally manage compute, storage, and cyber-physical resources with the objective of minimizing the total energy dissipation while meeting SLAs.

Finally, to evaluate datacenter designs, one must develop new methodologies and simulation infrastructure to quantify the impact and prototype research ideas. Because of the complexity and scale of datacenter applications, conventional evaluation approaches cannot evaluate new innovations with reasonable turnaround time. Hence, we must design hierarchical models, which integrate the performance and energy estimates across detail and time granularities, and parallel cluster-on-a-cluster simulation techniques, which together allow us to quantitatively evaluate systems at an entirely new scale.

REFERENCES

[1] Datacenter Dynamics Global Industry Census 2011 (Online). Available from: http://www.datacenterdynamics.com.br/research/market-growth-2011-2012.

[2] G. Cook, How Clean is Your Cloud? Catalysing an Energy Revolution, Greenpeace International, Amsterdam, The Netherlands, 2012.

[3] ENERGY STAR, Report to Congress on Server and Datacenter Energy Efficiency Public Law 109-431, U.S. Environmental Protection Agency, Washington, DC, 2007.

[4] D. Meisner, B. Gold, T. Wenisch, PowerNap: eliminating server idle power, in: Proceedings of the ACM International Conference on Architectural Support for Programming Languages and Operating Systems, Washington, DC, 2009.

[5] S. Pelley, D. Meisner, T.F. Wenisch, J. VanGilder, Understanding and abstracting total datacenter power, in: Workshop on Energy-Efficient Design, 2009.

[6] EPA Conference on Enterprise Servers and Datacenters: Opportunities for Energy Efficiency, EPA, Lawrence Berkeley National Laboratory (2006).

[7] L.A. Barroso, U. Hölzle, The case for energy-proportional computing, IEEE Computer 40 (2007) 33–37.

[8] L.A. Barroso, U. Holzle, The Datacenter as a Computer: An Introduction to the Design of Warehouse-Scale Machines, Morgan & Claypool Publishers, California, USA, 2009.

[9] P. Barham, B. Dragovic, K. Fraser, S. Hand, T. Harris, A. Ho, R. Neugebauer, I. Pratt, A. Warfield, Xen and the art of virtualization, in: 19th ACM Symposium on Operating Systems Principles, 2003.

[10] M. Armbrust, A. Fox, R. Griffith, A.D. Joseph, R. Katz, A. Konwinsk, G. Lee, D. Patterson, A. Rabkin, I. Stoica, M. Zaharia, A view of cloud computing, Commun. ACM 53 (4) (2010) 50–58.

[11] R. Buyya, Market-oriented cloud computing: vision, hype, and reality of delivering computing as the 5th utility, in: 9th IEEE/ACM International Symposium on Cluster Computing and the Grid, CCGRID, 2009.

[12] A. Karve, T. Kimbre, G. Pacifici, M. Spreitzer, M. Steinder, M. Sviridenko, A. Tantawi, Dynamic placement for clustered web applications, in: 15th International Conference on World Wide Web, WWW'06, 2006.

[13] C. Tang, M. Steinder, M. Spreitzer, G. Pacifici, A scalable application placement controller for enterprise datacenters, in: 16th International World Wide Web Conference, WWW2007, 2007.

[14] F. Chang, J. Ren, R. Viswanathan, Optimal resource allocation in clouds, in: 3rd IEEE International Conference on Cloud Computing, CLOUD 2010, 2010.

[15] J.S. Chase, D.C. Anderson, P.N. Thakar, A.M. Vahdat, R.P. Doyle, Managing energy and server resources in hosting centers, in: 18th ACM Symposium on Operating Systems Principles (SOSP'01), 2001.

[16] E. Pakbaznia, M. GhasemAzar, M. Pedram, Minimizing datacenter cooling and server power costs, in: Proceedings of Design Automation and Test in Europe, 2010.

[17] S. Srikantaiah, A. Kansal, F. Zhao, Energy aware consolidation for cloud computing, in: Conference on Power Aware Computing and Systems (HotPower'08), 2008.

[18] J. Kim, M. Ruggiero, D. Atienza, M. Lederberger, Correlation-aware virtual machine allocation for energy-efficient datacenters, in: Proceedings of the Conference on Design, Automation and Test in Europe, 2013.

[19] I. Hwang, M. Pedram, Portfolio theory-based resource assignment in a cloud computing system, in: IEEE 5th International Conference on Cloud Computing (CLOUD), 2012.

[20] A. Corradi, M. Fanelli, L. Foschini, VM consolidation: a real case based on OpenStack cloud, Futur. Gener. Comput. Syst. 32 (2014) 118–127.

[21] Z. Xiao, W. Song, Q. Chen, Dynamic resource allocation using virtual machines for cloud computing environment, IEEE Trans. Parallel Distrib. Syst. 24 (6) (2013) 1107–1117.

[22] D. Novakovic, N. Vasic, S. Novakovic, D. Kostic, R. Bianchini, DeepDive: Transparently Identifying and Managing Performance Interference in Virtualized Environments, EPFL, Lausanne, Switzerland, 2013.

[23] A. Verma, P. Ahuja, A. Neogi, pMapper: power and migration cost aware application placement in virtualized systems, in: ACM/IFIP/USENIX 9th International Middleware Conference, 2008.

[24] S. Martello, P. Toth, Knapsack Problems: Algorithms and Computer Implementations, Wiley, New Jersey, USA, 1990.

[25] S. Takeda, T. Takemura, A rank-based VM consolidation method for power saving in datacenters, Inf. Media Technol. 5 (3) (2010) 994–1002.

[26] H. Goudarzi, M. Pedram, Maximizing profit in the cloud computing system via resource allocation, in: Proceedings of International Workshop on Datacenter Performance, 2011.

[27] B. Urgaonkar, P. Shenoy, T. Roscoe, Resource overbooking and application profiling in shared hosting platforms, in: Symposium on Operating Systems Design and Implementation, 2002.

[28] Z. Liu, M.S. Squillante, J.L. Wolf, On maximizing service-level-agreement profits, in: Third ACM Conference on Electronic Commerce, 2001.

[29] K. Le, R. Bianchini, T.D. Nguyen, O. Bilgir, M. Martonosi, Capping the brown energy consumption of internet services at low cost, in: International Conference on Green Computing (Green Comp), 2010.

[30] L. Zhang, D. Ardagna, SLA based profit optimization in autonomic computing systems, in: Proceedings of the Second International Conference on Service Oriented Computing, 2004.

[31] D. Ardagna, M. Trubian, L. Zhang, SLA based resource allocation policies in autonomic environments, J. Parallel Distrib. Comput. 67 (3) (2007) 259–270.

[32] D. Ardagna, B. Panicucci, M. Trubian, L. Zhang, Energy-aware autonomic resource allocation in multi-tier virtualized environments, IEEE Trans. Serv. Comput. 99 (2010) 2–19.

[33] H. Goudarzi, M. Pedram, Multi-dimensional SLA-based resource allocation for multi-tier cloud computing systems, in: Proceeding of 4th IEEE Conference on Cloud Computing (Cloud 2011), 2011.

[34] G. Tesauro, N.K. Jong, R. Das, M.N. Bennani, A hybrid reinforcement learning approach to autonomic resource allocation, in: Proceedings of International Conference on Autonomic Computing (ICAC'06), 2006.

[35] D. Kusic, J.O. Kephart, J.E. Hanson, N. Kandasamy, G. Jiang, Power and performance management of virtualized computing environments via lookahead control, in: Proceedings of International Conference on Autonomic Computing (ICAC'08), 2008.

[36] H. Goudarzi, M. Ghasemazar, M. Pedram, SLA-based optimization of power and migration cost in cloud computing, in: 12th IEEE/ACM International Conference on Cluster, Cloud and Grid Computing (CCGrid), 2012.

[37] E. Feller, C. Morin, A. Esnault, A case for fully decentralized dynamic VM consolidation in clouds, in: IEEE 4th International Conference on Cloud Computing Technology and Science (CloudCom), 2012.

[38] C. Mastroianni, M. Meo, G. Papuzzo, Probabilistic consolidation of virtual machines in self-organizing cloud data centers, IEEE Trans. Cloud Comput. 1 (2) (2013) 215–228.

[39] M.N. Bennani, D.A. Menasce, Resource allocation for autonomic datacenters using analytic performance models, in: Second International Conference on Autonomic Computing, 2005.

[40] B. Urgaonkar, G. Pacifici, P. Shenoy, M. Spreitzer, A. Tantawi, An analytical model for multi-tier internet services and its applications, in: SIGMETRICS 2005: International Conference on Measurement and Modeling of Computer Systems, 2005.

[41] M. Pedram, I. Hwang, Power and performance modeling in a virtualized server system, in: 39th International Conference on Parallel Processing Workshops (ICPPW), 2010.

[42] A. Chandra, W. Gongt, P. Shenoy, Dynamic resource allocation for shared datacenters using online measurements, in: International Conference on Measurement and Modeling of Computer Systems ACM SIGMETRICS, 2003.

[43] N. Bobroff, A. Kochut, K. Beaty, Dynamic placement of virtual machines for managing SLA violations, in: Proceedings of the 10th IFIP/IEEE International Symposium on Integrated Management (IM2007), 2007.

[44] A. Beloglazov, R. Buyya, Adaptive threshold-based approach for energy-efficient consolidation of virtual machines in cloud data centers, in: Proceedings of the 8th International Workshop on Middleware for Grids, Clouds and e-Science, 2010.

[45] A. Beloglazov, R. Buyya, Optimal online deterministic algorithms and adaptive heuristics for energy and performance efficient dynamic consolidation of virtual machines in cloud data centers, Concurrency and Computation: Practice and Experience (2012) 1397–1420.

[46] M. Wang, X. Meng, L. Zhang, Consolidating virtual machines with dynamic bandwidth demand in data centers, in: IEEE INFOCOM, 2011.

[47] T. Hirofuchi, H. Nakada, S. Itoh, S. Sekiguchi, Reactive consolidation of virtual machines enabled by postcopy live migration, in: Proceedings of the 5th International Workshop on Virtualization Technologies in Distributed Computing, 2011.

[48] R. Raghavendra, P. Ranganathan, V. Talwar, Z. Wang, X. Zhu, No "power" struggles: coordinated multi-level power management for the datacenter, ACM SIGPLAN Not. 43 (3) (2008) 48–59.

[49] X. Fan, W. Weber, L.A. Barroso, Power provisioning for a warehouse-sized computer, in: Proceedings of the 34th Annual International symposium on Computer Architecture, San Diego, CA, 2007.

[50] S. Pelley, D. Meisner, P. Zandevakili, T.F. Wenisch, J. Underwood, Power routing: dynamic power provisioning in the datacenter, in: ASPLOS'10: Architectural Support for Programming Languages and Operating Systems, 2010.

[51] A. Gandhi, M. Harchol-Balter, R. Das, C. Lefurgy, Optimal power allocation in server farms, in: International Joint Conference on Measurement and Modeling of Computer Systems (SIGMETRICS'09), 2009.

[52] W. Felter, K. Rajamani, T. Keller, C. Rusu, A performance-conserving approach for reducing peak power consumption in server systems, in: 19th Annual International Conference on Supercomputing (ICS'05), 2005.

[53] M. Srivastava, A. Chandrakasan, R. Brodersen, Predictive system shutdown and other architectural techniques for energy efficient programmable computation, IEEE Trans. VLSI 4 (1996) 42–55.

[54] Q. Qiu, M. Pedram, Dynamic power management based on continuous-time Markov decision processes, in: ACM Design Automation Conference (DAC'99), 1999.

[55] G. Dhiman, T.S. Rosing, Dynamic power management using machine learning, in: ICCAD'06, 2006.

[56] E. Elnozahy, M. Kistler, R. Rajamony, Energy-efficient server clusters, in: Proceedings of 2nd Workshop Power-Aware Computing Systems, 2003.

[57] D. Meisner, C. Sadler, L. Barroso, W. Weber, T. Wenisch, Power management of online data-intensive services, in: Proceedings of the 38th Annual International symposium on Computer Architecture, 2011.

[58] X. Wang, Y. Wang, Co-Con: coordinated control of power and application performance for virtualized server clusters, in: IEEE 17th International Workshop on Quality of Service (IWQoS), 2009.

[59] R. Buyya, A. Beloglazov, Energy efficient resource management in virtualized cloud datacenters, in: 10th IEEE/ACM International Conference on Cluster, Cloud and Grid Computing (CCGrid), 2010.

[60] Google Cloud Platform Trace (Online). Available from: https://cloud.google.com/tools/cloud-trace.

[61] http://ark.intel.com/ (online).

[62] http://aws.amazon.com/ec2/#pricing (online).

ABOUT THE AUTHORS

Hadi Goudarzi received a BSc and MSc degree from the Sharif University of Technology, Tehran, Iran, in 2006 and 2008, respectively, both in electrical engineering (communications), and a PhD degree in electrical engineering from the University of Southern California in 2013. Since 2013, he has been with Qualcomm, Inc., where he is currently a staff engineer. His current research interests include energy-efficient computing, system-level low-power design, and optimization.

Massoud Pedram received a BS degree in EE from the California Institute of Technology in 1986 and a PhD degree in EECS from the University of California, Berkeley in 1991. He then joined the EE department of USC where he is currently a professor. He has published more than 500 journal and conference papers, written 4 books on various aspects of low-power design, and holds 10 U.S. patents. His research has received a number of awards including two ICCD Best Papers, two DAC Best Papers, and an IEEE T-VLSI Best Paper. He is a recipient of the NSF's Young Investigator Award (1994) and the Presidential Faculty Fellows Award (a.k.a. PECASE Award) (1996). His current research interests include energy-efficient computing, energy storage systems, low-power electronics and design, and computer-aided design of very large-scale integration circuits and systems. Dr. Pedram is an IEEE Fellow and ACM Distinguished Scientist.

CHAPTER FIVE

Communication-Awareness for Energy-Efficiency in Datacenters

Seyed Morteza Nabavinejad, Maziar Goudarzi
Department of Computer Engineering, Energy Aware Systems Lab, Sharif University of Technology, Tehran, Iran

Contents

Abstract

With the proliferation of cloud computing concept, the datacenters, as the basic infrastructure for cloud computing, have gained an ever-growing attention during the last decade. Energy consumption in datacenters is one of the several features of them that have been the target of various researches. Two major consumers of energy in datacenters are the cooling system and IT equipment. Computing resources, such as servers, and communicating ones, such as switches, constitute the main portion of IT equipment. Among these two major players, the servers have been considered more than networking equipment. Making servers energy proportional as well as server

Advances in Computers, Volume 100
ISSN 0065-2458
http://dx.doi.org/10.1016/bs.adcom.2015.11.008

consolidation are the two essential approaches regarding reduction of servers' energy consumption. However, some researches indicate that 10–20% of energy consumption of IT equipment goes to network equipment and hence they must also be considered en route to better energy consumption in datacenters. The focus of this chapter is energy consumption of network equipment in datacenters and conducted researches in this area. First, a quick summary about network energy consumption in datacenters is presented. After that, related state of the art approaches and techniques are categorized, reviewed, and discussed. Finally, the chapter is concluded with presentation of recent original work of authors and its details.

ABBREVIATIONS

BCN bidimensional compound network
CAVMP communication-aware VM placement
CIVSched communication-aware inter-VM scheduling
CS Communication skeleton
FFLM first-fit virtual link mapping
HCN hierarchical irregular compound network
ILP integer linear programming
IRA integrated resource allocator
IT information technology
MPI message-passing interface
NIC network-interface card
NPE network power effectiveness
NS2 network simulator 2
PM physical machine
PUE power usage effectiveness
QoS quality of service
SaaS storage as a service
SABVMP simulated annealing-based VM placement
SCAVP structural constraint-aware virtual machine placement
SDN software-defined network
TEA traffic-aware embedding algorithm
TOP-VCM topology-aware partial VC mapping
VC virtual cluster
VDC virtual datacenter
VM virtual machine
VMM virtual machine manager

1. INTRODUCTION

Energy consumption is a critical concern for operators of datacenters and there are several factors that affect this energy consumption. Various parts such as information technology (IT) equipment, cooling, and power distribution consume power in datacenters. One measure of power efficiency in datacenters is the power usage effectiveness (PUE) metric which is the ratio of total power consumption over IT equipment power consumption [1]. Many works have traditionally reduced the PUE by mostly considering the cooling power consumption [2–6].These efforts have led to very notable PUEs such as 1.12 for Google [1] or 1.07 for Facebook [7] by reducing total power consumption in the PUE definition. More recent works focus on improving power consumption of the IT equipment. Between severs and switches which are the major IT equipment that consume power, many works have been done to reduce the energy consumption of servers so as to make them energy proportional [8–11]. A study [12] on various kinds of Web servers shows that the average utilization of servers varies between 11% and 50%. Virtualization has improved utilization of today datacenters by allowing consolidating several virtual machines (VMs) on a single physical machine (PM) server. Since most today servers are not energy proportional [13], this consolidation helps operate the servers in their more energy-efficient operating regions. Virtualized datacenters enable a datacenter to service more requests per unit time and energy compared to nonvirtualized ones which results in a greener datacenter with smaller carbon footprint. The amount of energy reduction by VM consolidation highly depends on the VM placement algorithm employed in the datacenter. Many different VM consolidation techniques have been presented [14–21] and each of them considers different parameters to reduce the number of ON PMs and to place VMs on them.

Aforementioned references indicate the importance of servers in datacenters; however, measurement reports reveal that about 10–20% of power consumption of IT equipment goes to network equipment [22]. A case study on Google datacenters also indicates that when the server utilization is 100%, the network infrastructure consumes around 20% of total IT equipment power, but when the server utilization drops down to 15%, the network share increases to 50% and even higher when energy-proportional servers are employed [23] (see Fig. 1). These observations demonstrate the significance of network power consumption and its need for more attention.

This chapter surveys various aspects of network power consumption in datacenters and reviews state of the art approaches and mechanisms. In Section 2, power consumption of components that constitute the datacenter networks are presented and discussed. Section 3 is dedicated to categorization of numerous techniques that are presented for energy

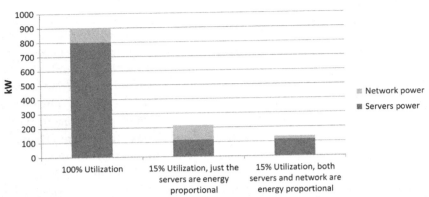

Figure 1 Impact of energy proportionality of network on total power consumption [23].

reduction of network in datacenters. The remaining sections of chapter are dedicated to the recent approach of authors toward reducing energy consumption of datacenters with respect to communication.

2. POWER CONSUMING COMPONENTS IN NETWORKS

There are two parts in the networking infrastructure that consume power: switches and the network–interface cards (NICs) of servers. According to Ref. [24], NICs consume only about 5% of total power in servers; thus, the network switches are the main target when addressing power consumption in networks. In Fig. 2, the share of each component in total power consumption of IT equipment is demonstrated. As can be seen, the share of switches is way more than NICs in servers.

For a better understanding of the power consumption mechanism in network switches, note that each switch consists of a chassis containing a mainboard with one or more linecards connected to it and each linecard has several ports. The power consumption of ports varies with their bit rate. Depending on the equipment, bit rate of ports can be individually set and they can even be turned off if not connected or not needed. Similarly, each linecard can be separately turned on or off. In general, we can say that Energy consumption of a switch depends on the following factors: (a) type of switch, (b) number of ports, (c) port transmission rates, and (d) employed cabling solutions [19]. Thus, total power consumption of a network switch can be formulated as Eq. (1) [25]:

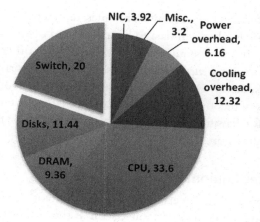

Figure 2 Distribution of power consumption in IT equipment [24].

$$P_{\text{switch}} = P_{\text{chassis}} + n_{\text{linecard}} * P_{\text{linecard}} + \sum_{i=0}^{R} n_{\text{ports}} * P_r \qquad (1)$$

where P_{chassis} is the amount of power the switch consumes when turned ON, regardless of other parameters. The n_{linecard} parameter indicates the number of linecards in chassis and P_{linecard} is the power consumption of an active linecard. Finally, n_{ports} is the number of ports in switch and P_r corresponds to power consumed by an active port (transmitter) running at the rate r.

3. ENERGY REDUCTION TECHNIQUES

As Eq. (1) shows, power consumption of network switches can be reduced in several ways.

For classification purposes, we divide these approaches into three categories: use better equipment, better use of equipment, and reduce use of equipment. The first category deals with ways to reduce the constant values in Eq. (1) so that the network equipment is more energy proportional, i.e., their power consumption is effectively reduced to near zero when not in use. The second category, *better use of equipment*, deals with static and dynamic techniques that help turn-off more linecards to save energy. These two categories do not affect total communication volume transmitted over the network, but try to do it in a more efficient way by actually reducing *power* consumption of the equipment. The third category, which comprises more sophisticated and more recent techniques, represents various approaches that influence total volume of data to be transmitted by the network so that total operating time of equipment is reduced. In other words, these techniques reduce total *energy* consumption of the network equipment by reducing the *time* factor. Obviously, one needs to simultaneously apply these three categories of techniques to obtain the most effective energy reduction outcomes, but for classification purposes, we present each of them separately in the following section.

3.1 Use Better Equipment

Equation (1) practically covers most popular network switches in widespread use today. The constant values, such as the power consumption of the mainboard or the base power of each linecard, plays an important role in total power consumption; even when there is no activity in the network, these constant values are consumed to operate the switch. Reducing these constants is one important way to make switches energy proportional. Moreover, the channels in many current switches are always ON regardless of transmission rate (even when there is no packet to transmit). More advanced switches, such as Dynamic InfiniBand switches, can be adapted to transmission rate and save power [26–28]. Use of other technologies, such as optical cables and networks, is another way for energy proportionality [29–31]. For example, above InfiniBand switches along with optical links make it possible for links to operate with fewer lanes and at a lower data rate to reduce the power consumption [23].

3.2 Better Use of Equipment

In this category, the components of a switch such as ports or linecards are considered and the main concern is to reduce the power consumption of them in order to reduce the energy consumption of switches, and consequently, reducing the network energy consumption. For example, in Ref. [32], the behavior of ports in switches is studied and a number of techniques are proposed as below: the simplest technique suggests disabling unused ports since some of them are never used or used just for short periods of time. This technique is inexpensive to implement statically, but if the communication requirements change dynamically, turning on–off should also be dynamic which is harder to implement. Port rate adaptation is another technique they suggest that can be useful for ports with low and almost constant utilization, but not for ones with fluctuating utilization. The third technique proposes aggregating the entire set of active ports on as few linecards as possible. It makes it possible to shut down the unused linecards to save the energy. Finally, the linecards can also similarly be consolidated across fewer switches. It provides the opportunity to reduce the number of active switches and save energy. These works [33–37] have used one or more of aforementioned techniques. However, moving ports from sparse switches to crowded ones needs significantly rewiring the network.

3.3 Reducing the Usage of Equipment

These techniques are more focused on concepts such as communication time, communication volume, or communication scheduling in datacenter network. Each of the aforementioned concepts can affect the energy consumption of datacenter network. For example, if one can decrease the amount of data that transfers among the network, the usage of switches will decrease and consequently they consume less energy. In the following sections, we will categorize these kinds of approaches and give more examples and explanation about each subcategory. In Fig. 3, you can see an overview of subcategories.

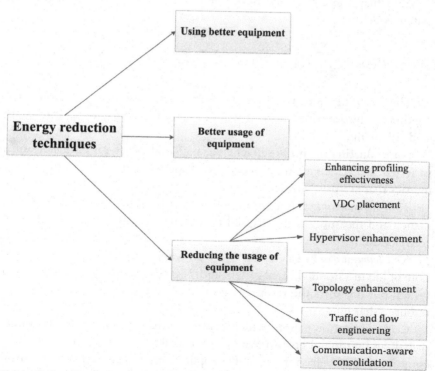

Figure 3 Overview of energy reduction techniques.

3.3.1 Enhancing Profiling Effectiveness

These kinds of works are the first ones in the area of communication-aware task placement algorithms and are the base for many algorithms that are proposed for datacenters and clouds. Reference [38] is one of the primitive works that considers the communication when trying to balance the load in a cluster that runs parallel workloads. COM-aware is designed to handle a wide variety of applications by means of an application behavioral model that includes load of CPU, disk, and network generated by application.

For constructing this model, a profiling step is needed. The profiling can be either offline or online. This model then will be used to calculate the load impact of each application on cluster. The model considers several phases for a process of application and for each phase, the total execution time of that phase is broken down to communication time, disk I/O time, and CPU time. Below equation shows this break down by this assumption that there are N phases:

$$\text{for each } i \text{ in } [1...N]: \quad T^i = T^i_{\text{COM}} + T^i_{\text{Disk}} + T^i_{\text{CPU}}.$$

After that, the communication, ddisk, and CPU requirements are estimated as below:

$$R_{\text{COM}} = \sum_{i=1}^{N} T^i_{\text{COM}}, \quad R_{\text{Disk}} = \sum_{i=1}^{N} T^i_{\text{Disk}}, \quad R_{\text{CPU}} = \sum_{i=1}^{N} T^i_{\text{CPU}}.$$

The profiling step mentioned earlier is responsible for obtaining the T^i_X times. This information then will be used to do communication-aware load balancing.

For evaluating the proposed schema, a 32-node cluster is simulated and a variety of workloads are used to evaluate its performance. The performance metrics that are used for evaluating are

Turn-around time: it is a common metric for evaluating the job performance and measures as the elapsed time between job submission and job completion.

Slowdown: it can be calculated as T_p'/T_p where the T_p' is the turn-around time of job in a nondedicated system and T_p is turn-around time in the same system but this time the resources are dedicated to the job and there is no resource sharing.

Communication pattern which consists of three key attributes (volume, spatial, and temporal characteristics) is a useful resource for perceiving the communication behavior of parallel applications. It can be extracted from communication trace. Before [39], the parallel applications had to run on whole cluster in order to produce communication trace but this process has two main weaknesses: it uses lots of resources and takes a long time. This work proposes a new method for producing communication trace by less resources and time, but by sacrificing the temporal attribute of communication pattern. It uses message-passing interface (MPI) program and analyzes its messages to find the critical parts and then uses them to generate reliable communication trace. This work is also one of the first works that are done for communication awareness and later are applied to clouds and datacenters.

3.3.2 VDC Placement

For service providers who deploy their services on the cloud, sole VMs without considering their communication overhead cannot guarantee the desired performance. Instead they expect an entity that considers computing and communication elements simultaneously. For satisfying such demands, the virtual datacenter (VDC) concept emerged. A VDC is a set of VMs, switches, and routers that are connected through virtual links and each virtual link is characterized by its bandwidth and delay. The VDCs can address the performance issue since they are kind of isolated regarding network and hence can better guarantee network resources availability.

Mapping the VDC resources onto physical infrastructure resources efficiently is a problem that several works are proposed to address it. Reference [40] proposes an approach for placing VDCs that aims to achieve two goals: allocating the computing and networking resources to each VDC and maximizing the number of placed VDCs and guaranteeing their performance. Since the available bandwidth for each VDC can affect the performance of running tasks on it, it is important to consider the network bandwidth in VDC placement. This work tries to reduce the inter-VDC bandwidth in order to increase the performance. It first proves that the problem of clustering the VMs of a VDC in order to reduce the inter-cluster bandwidth is NP-hard. After that, the three-phase integrated resource allocator, IRA, is introduced that exploits the min-cut algorithms to form the VM-clusters. Finally, another version of IRA, called B-IRA, is proposed that uses approximation algorithms and explores a smaller space for solution. Three phases of IRA can be seen in Fig. 4.

The *first phase* of IRA is only applied when the resource requirement of VDC is more than one server-cluster. All the servers in a server-cluster are attached to the same edge-switch. Therefore, when more than one server-cluster is required it means that there will be communication between

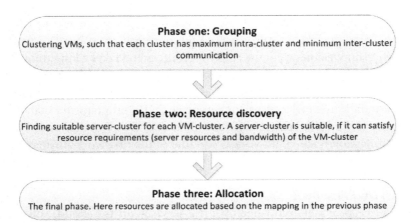

Figure 4 Three phases of IRA.

clusters. Because of the high cost of this kind of communication compared with intra-cluster one, the IRA tries to cluster the VMs of VDC in a way that inter-cluster is as low as possible. It is obvious that when the VDC can be fit in a single server-cluster, there is no need for VM-clustering. The *second phase* tries to find a valid mapping for each VDC on physical resources. The process of mapping for each VDC is successful if: first, there is at least one sever-cluster for each VM-cluster of VDC that satisfies the resource demand of the VM-cluster, and second, there is enough bandwidth among the selected server-clusters that can handle the communication between VM-clusters. Finally in the *third phase*, when the two previous phases are done successfully, the allocation routine places the VM-clusters onto corresponding server-clusters.

For evaluating the proposed technique, a set of simulations is conducted. The physical infrastructure considered is a three-tier hierarchical architecture with 192 server-clusters and 9216 servers. Regarding the number of switches, there are 8 core switches, 32 aggregation switches, and 192 edge-switches, and each edge-switch represents a server-cluster. For evaluating the performance, three metrics are selected:

Percentage of accepted VDCs: how many of requested VDCs are accepted and placed on infrastructure?

Cost: here the cost is the inter-cluster communication. When a VDC is split into several VM-clusters, the amount of communication between these parts is the cost of VDC placement.

Partition size: by changing the size of VDCs, the effect of VM-clustering is observed.

The main assumption in Ref. [41] is that network switches are the bottle-neck in datacenter network and that the datacenter network topology is fat-tree. They have designed two algorithms called traffic-aware embed-ding algorithm (TAE) and first-fit virtual link mapping (FFLM) to map the requests for VDC on the physical infrastructure. They claim that pro-posed approach can increase the chance of placing VDCs and reduce the network cost.

Virtual clusters (VC) are entities similar to VDCs that provide isolated resources for a job and are configured per job in runtime. A job that runs on VC has several sub-jobs and each of them runs on a VM in VC. The proposed approach in Ref. [42] first extracts the communication pattern among the VMs in VC and makes communication skeleton (CS). It then tries to place the VC on infrastructure by considering the resource demand of VMs as well as derived CS in order to increase resource (CPU and network) utilization. For reaching this goal, they introduce partial VC mapping which is in contrast to full VC mapping. They claim the former can lead to better resource utilization and less placement com-plexity. topology-aware partial VC mapping (TOP-VCM) algorithm is designed to fulfill this concept. TOP-VCM tries to map VMs and virtual links between them on physical resources and links in order to increase resource utilization. It also leads to decrease in response time of tasks which are processed by job that is running on VC due to better communication performance.

The majority of works done have considered only the case where there is one infrastructure but distributed infrastructures are not considered. So Ref. [43] tries to solve the VDC embedding problem in the case of distributed infrastructures by considering energy efficiency and environmental impacts. The proposed solution comprises two phases: *VDC partitioning* and *partition embedding*. VDC partitioning phase splits the VDC to several partitions such that bandwidth between partitions is minimized. The partition embedding phase then lists the datacenters that are able to host the partition and chooses the best one based on constraints such as cost and location.

3.3.3 Hypervisor Enhancement
Virtualization technology has a promising effect on resource utilization in datacenters. Since the main concern at the time of designing current hyp-ervisors such as Xen was the computing-intensive applications they suffer from poor network I/O performance. For example, single-root I/O virtualization, which is the current standard for network virtualization, is

based on interrupt. Since handling each interrupt is costly, the performance of network virtualization depends on the resource allocation policy in each hypervisor.

For conquering aforementioned problem, Ref. [44] proposes a packet aggregation mechanism that can handle the transfer process in a more efficient and rapid manner. However, the aggregation step itself introduces a new source of delay and must be addressed. Hence, the queuing theory has been used to model and dynamically tune the system in order to achieve the best tradeoff between delay and throughput.

Regarding the resource allocation policy problem, the credit-based scheduler, the *de facto* resource scheduler in Xen hypervisor, cannot handle the I/O-intensive workloads properly. The reason is that it is not aware of different behaviors of various VMs and handles all of them in the same way. So, the I/O-intensive VMs do not earn enough credit for handling network interrupts and hence retrieve the data in a slow manner that leads to high latency and response time.

To tackle the mentioned problem and eliminate the bottleneck which is caused by scheduler, Ref. [45] introduces a *workload-aware network virtualization model*. This model monitors the behavior of VMs and divides them based on their behavior in two categories: *I/O-intensive* and *CPU-intensive* and then handles them by *Shared Scheduling* and *Agile Credit Allocation*. When an I/O-intensive VM faces burst traffic, the shared scheduling gives it more credit so it could be able to handle the traffic. Agile credit allocation is responsible for adjusting the total credit based on number of I/O-intensive VMs in order to reduce the wait time for each I/O-intensive VM.

Reference [46] first introduces a semantic gap that exists between VMM and VMs. The gap is that VMM is unaware of processes inside VMs so it cannot schedule the VMs efficiently. As an example, when a VM sends a request to another co-located VM, the co-located VM must earn vCPU so it can process the request and provide the response ASAP but the current scheduling in VMM does not consider this matter. As a result, the response latency increases and it may lead to quality of service (QoS) violation. Moreover, it gets worsen when the co-located VMs are CPU or IO intensive because the competition for CPU increases dramatically. Figure 5 illustrates the mechanism of inter-VM communication in current hypervisors.

To resolve the problem, they propose the communication-aware inter-VM scheduling (*CIVSched*) algorithm which is aware of communication among co-located VMs. CIVSched monitors the packets that are send

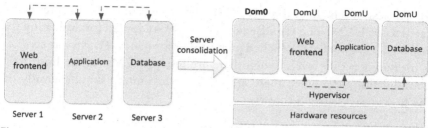

Figure 5 Inter-VM communication in VMMs.

through the network and identifies the target VM and schedules the VM in a way to reduce response latency. The CIVSched prototype is implemented on Xen hypervisor.

For each DomU guest (VM), there is a virtual front-end driver that VM sends the requests for I/O operations to it. Then these requests are sent to back-end driver which is in Dom0 guest. And finally, the back-end driver sends the captured requests to real device driver and returns the responses to the front-end driver.

The CIVSched must abide by two design principles: low latency for inter-VM and low latency for the inner-VM process. These two design principles help CIVSched to decrease the inter-VM latency. For realizing the two above-mentioned requirements, the CIVSched adds five modules to the Xen I/O mechanism. The *AutoCover* (Automatic Discovery) module finds the co-located VMs and stores their MAC address and IDs in a mapping table. The *CivMonitor* checks all the packets transmitted by VMs and when finds an inter-VM packet, informs the *CivScheduler* about it. Then, *CivScheduler* gives more credit to target VM so it can handle the packet as fast as possible. Until now, the first design principle (low latency for inter-VM) is satisfied but the other one still needs attention. Regarding the second principle, *CivMonitor* identifies the process of target VM that will receive the packet via TCP/UDP port number within the packet and passes the information to the target VM. Finally, *PidScheduler* and *PidTrans* modules inside the guest VM schedule the target process with respect to decreasing latency.

For evaluating the CIVSched, it has been implemented on Xen hypervisor version 4.1.2 and is compared with XenandCo [47] scheduler (another proposed scheduler for Xen) and Credit scheduler which is the base scheduler in Xen. For comparing the *Network latency*, experiments consist of a ping-pong test, a simulation test and a real-world Web application scenario but with synthetic benchmarks. *Fairness Guarantees* is also evaluated because the fairness of scheduler directly affects the fairness of CPU resources allocated to each VM. The UnixBench suite 4.1.0 is adopted for evaluating the *performance overhead* of CIVSched on host's performance. *Performance overhead* is measured at two levels: when there are just two VMs on the host (light consolidation) and when there are seven VMs running simultaneously on the host (heavy consolidation).

3.3.4 Topology Enhancement

There are two major categories for datacenters network architecture: *switch-centric* and *server-centric*. In *server-centric architectures*, the servers not only act as computing resources, but also have the responsibility for packet forwarding. In *switch-centric architectures*, however, the packet forwarding and communication is guaranteed by switches and servers do not have any role in packet delivery.

Among the switch-centric architectures, three-level fat-tree topology is the most common one. In this topology, there are several paths between two nodes so in the case of link failure there are alternative routing ways. One dominant problem by fat-tree is that the number of nodes that can be connected is restricted. If one needs to connect more nodes, there are two ways for that: replacing the switches with bigger ones that need to reconfigure the whole connections which is costly or increasing the number of levels that leads to increase in network diameter. For solving the mentioned problems, Ref. [48] proposes a new approach for constructing large datacenter networks with fixed-size switches. They first construct hypergraphs using the hypergraph theory and then convert them to indirect hypergraphs. This approach makes it possible to connect more nodes together with fixed-size switches compared with fat-tree topology.

In server-centric architectures, the delay of a server-to-server direct hop and a server-to-server-via-a-switch hop is considered equal; however, with the fast growing capabilities of servers for packet forwarding, this assumption might be invalid in the future. Moreover, until today, it is believed that bidimensional compound network (BCN) [49] can connect the most number of dual-port servers for a fixed number and size of switches, while the authors claim that DPillar [50] architecture can connect more servers than BCN under the same configuration.

Regarding two aforementioned points, Ref. [51] proposes three new server-centric architectures for dual-port servers. The architectures try to answer this question: "what is the maximum number of dual-port servers that any architecture can accommodate at most, given network diameter *d*, and switch port number *n*" [51] and approximate the upper bound of possible dual-port servers.

The first architecture, based on generalized hypercube [52], is called SWCube. Under different conditions, this architecture either can connect more servers than DPrill or less. The two other architectures, SWKautz and SWdBrujin, which are based on Kautz graph [53] and de Brujin graph [54], respectively, always provide better answers compared with DPrill.

With the help of dual-port servers, Ref. [49] proposes two new server-centric network structures called hierarchical irregular compound network (HCN) and BCN for datacenters which are of server degree 2. These structures have low diameter and high bisection and can be expanded easily.

A level-*h* HCN is denoted as HCN(*n,h*). Here, *n* stands for number of dual-port servers in the smallest module and number of ports in miniswitch that connects the servers together in that module. A HCN(*n,h*) consists of *n* of HCN(*n,h*−1) modules that are connected via a complete graph. Here the second port of servers is used to connect the smallest modules together (the first port is connected to miniswitch). The smallest module is denoted as HCN(*n*,0). Generally, we can say that

$$\text{for } i \geq 0, \quad \text{HCN}(n, i) \text{ is formed by } n\text{HCN}(n, i-1)$$

α stands for number of master servers and β determines the number of slave servers in BCN(α,β,h) where *h* shows the level of BCN in the first dimension. The sum of α and β denotes the number of dual-port servers and number of ports in miniswitch again in smallest module or building block similar to HCN. A general BCN is denoted by BCN(α,β,h,γ) and γ stands for level of BCN in the second dimension. It is worth mentioning that there is really no master/slave relation between servers and these names are just for simplifying the presentation of structure. The master servers are used to expand the BCN from outer section or first dimension and slave server make it possible to expand the BCN from inner section or second dimension.

Since good support for one-to-one traffic routing leads to good all-to-one and one-to-all support, two algorithms are presented for routing one-to-one traffic in BCN. First, the single path routing without failure is studied and then it has extended to parallel multipath routing. Finally, the failures are also considered and addressed by using multipath between servers. *Network order* (number of servers in network), *bisection width*, and *path diversity* are the parameters that BCN and HCN are compared with FiConn based on them.

The results show that BCN can surpass the FiConn and yield better performance regarding aforementioned parameters.

Network power effectiveness (NPE) indicates the tradeoff between power consumption and throughput of a network and also shows the bit-per-second per watt (bps per watt) for a network. This parameter is important in today datacenters since the main concern of recent proposed architectures for datacenter network is throughput and power consumption is rarely considered, however the network consumes a big portion of overall power in a typical datacenter. Reference [55] does a comprehensive study on dominant advanced network architectures in datacenters such as fat-tree and BCube regarding their NPE. It also studies the effect of different parameters such as traffic load, power-aware routing, traffic pattern, and topology size on the NPE and compares the switch-centric architectures (which that try to propose a new architecture based on switches or improve tree architecture) such as VL2 and fat-tree against server-centric architectures (that use servers with multiple NICs and each server is involved in the packet forwarding flow) such as DCell [56] and FiConn [57].

3.3.5 Traffic and Flow Engineering

Various methods are proposed that aim to reduce just energy consumption of network. Reference [58] aims to reduce power consumption of core network in distributed cloud for different kind of applications. For content delivery services, they design a mixed-integer linear programming model. Based on this model, they conclude that replicating popular content on different clouds can reduce power consumption significantly compared to centralized content delivery model because of power reduction of network switches. For storage as a service (SaaS), they suggest that migrating content based on its access frequency can be beneficial. And finally, for VM placement, they propose to break the large VMs into several smaller VMs and place them on different clouds to facilitate the access of users from different locations and reduce the power consumption of network.

Despite the common approaches that try to optimize the network performance and energy via sole traffic engineering, Ref. [59] tries to consider both traffic engineering as well as network features such as topology and end-to-end connectivity. The proposed approach deeply explores both application characteristics and network features and then based on these observations does the VM assignment. This assignment will lead to favorable conditions in datacenter network that will be used for next step which is traffic engineering.

One common way for reducing the power consumption of switches is to reduce the number of active switches by flow aggregation and then put the rest of them in "sleep on idle" state. Although the flow aggregation works fine for application-limited flows, where the amount of data that need to be transmitted is low and one network link can be shared between several flows to transmit their data, it cannot handle the network-limited flows such as MapReduce applications. In network-limited flows, the application produces lots of data in a short period of time and then the capacity of network as well as the number of flows that compete on the bottleneck link determines the throughput for each flow. Consequently, although the number of active switches is reduced by aggregation technique, their uptime increases due to limited capacity for each flow.

Reference [60] addresses above issue by modeling the problem and considering both reducing the number of active switches as well as deadline and size of each flow. They design *willow* which uses software-defined network (SDN) technique and schedules the flows in terms of energy consumption. For achieving online scheduling, a greedy approximation algorithm is presented. This algorithm tries to use all the idle ports in a switch and then reduce the number of active switches such that the running duration of network is not increased.

Example in Fig. 6 shows that current techniques for scheduling flows, that just consider either minimizing the number of active switches or maximizing the network throughput, are not suitable for network-limited flows regarding energy efficiency. In this example, there are two flows: f1 from server 1 to server 4 and f2 from server 3 to server 6. Here, the following assumptions are considered: size of each flow $= Z$, the capacity of each link $= B$, power consumption of each switch $= P$, and deadline of each flow $= (2 * Z)/B$.

The scheduling algorithm in (1A) aims to maximize the network throughput so routes the flows as is depicted. Here, four switches are active for Z/B time, so the total network energy consumption is $(4 * Z * P)/B$. Meanwhile, the scheduling algorithm in right side tries to minimize the number of active switches. The result is three active switches with $(2 * Z)/B$ uptime and hence the network energy consumption is $(6 * Z * P)/B$. As can be seen, while the second algorithm uses one less switch compared with the first one, it uses more energy than the first one. However, in this scenario both scheduling schemas meet the deadline.

In scenario B, it is assumed that links h–i and f–j are faulty and out of service. Again, the left side in Fig. 6 illustrates the path that flow scheduling

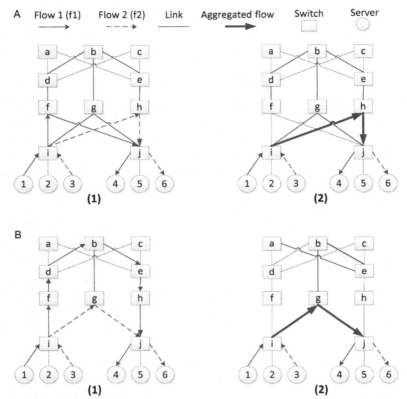

Figure 6 Illustrative example for showing performance of different flow scheduling algorithms.

maximizing throughput selects and the energy consumption of network for such routing would be $(8 * Z * P)/B$. The energy consumption of network for the path that the second algorithm, i.e., minimizing the active switches, chooses is $(6 * Z * P)/B$. We can see that despite the first scenario where the maximizing throughput gave better energy consumption, in this scenario minimizing the number of active switches is more successful regarding network energy consumption. This example clearly demonstrates that considering either network throughput or number of active switches is not sufficient and both of them must be considered simultaneously in order to achieve a successful scheduling under various conditions and the Willow algorithm is proposed to do it.

There are three basic ideas behind willow design and it has been developed based on them.

SDN-based flow scheduling: willow uses SDN framework to collect information about flows such as their size and deadline or computing the routing path for each of them.

Routing path selection: willow is not topology dependent and can work with different topologies such as fat-tree or Bcube.

Differentiation between elephant flows and mice flows: since the elephant flows dominate the traffic in datacenter networks [61], willow focuses on them and first schedules them, then reuses the computed paths for elephant flows in a random way for mice flows.

For evaluating Willow regarding network energy consumption reduction, simulation and testbed experiments are conducted. In the simulation setup, both fat-tree and blocking fat-tree are used for network topology. The MapReduce computation trace from 50 mappers and 20 reducers is considered for workload trace. The evaluation metric for comparing willow against rival algorithms, i.e., simulated annealing and particle swarm optimization, is network energy ratio. For evaluating willow in a real setup, a testbed with 16 servers in a fat-tree network is considered.

3.3.6 Communication-Aware Consolidation

As the amount of PUE goes down by using advanced approaches for reducing the non-IT equipment power consumption, the power consumption of IT equipment (servers and switches) becomes dominant in datacenters. A wide variety of techniques and algorithms are proposed for reducing the quota of servers in power consumption [14–16,20,62–64] but there are lots of opportunities for decreasing the switches' power consumption [60].

Joint approaches such as Refs. [65,66] consider both inter VMs' communication as well as VM consolidation. However, these works suffer from too simplistic and inaccurate models of communication time and its effect on application runtime and energy consumption. They have tried to group the VMs that communicate with one another and then place them on one or more servers to reduce network traffic, but did not take into account VMs communication inside a group, and more importantly, did not consider the communication structure among the servers.

To address the aforementioned issues regarding joint approaches, we have developed a new approach and will describe it here.

One of our contributions in this work is that we consider all inter-VM communication even inside a group and place VMs based on them so the network traffic is less compared to other approaches. Another advantage of our work is that we compute the communication time for each VM using a detailed network simulator, NS2, which improves the accuracy of results. A number of other communication-aware VM consolidation techniques exist, but they optimize only based on either the communication volume [67] or network congestion [19], both of which are relevant but indirect and inaccurate indicators of the actual communication time on the network. As an example, consider a unit volume of data communicated between two servers; it takes different amounts of time if the two servers are in the same rack compared to when they are in different racks, and even more if in different clusters. Thus, the distribution of communications endpoints as well as the structure of racks and clusters, in addition to communication volume, play important roles to determine communication time, and hence, communication energy consumption. We improve state of the art by showing this shortcoming in existing approaches and their resulting suboptimal placement, and by providing an approach to accurately consider network communication time.

An important contribution of our approach over prior works is that we consider the structure of the datacenter, in terms of placement of servers in racks as well as racks in clusters, when placing the VMs. Prior art have only concentrated on consolidating VMs on servers without paying attention to the reality that most, if not all, cloud computing implementations take advantage of a datacenter in the back-end, and hence, the actual place and connection structure of those servers is an important factor in total communication time, and the final energy consumption. We show and quantify the significance of this point by our experiments in Section 6.

Another contribution of ours is revealing the inaccuracies in prior work by considering actual network structure and elements when evaluating the energy consumption and VM consolidation outcome. To evaluate our as well as rival techniques, we use NS2 simulator for real-world network simulation, which is far more accurate than approaches that abstract out most network properties and only count the number of switches between servers, such as Ref. [68] where a switch can become hot spot, and consequently, the actual communication time is effectively more than they estimate. We prove superiority of our proposal by comprehensive experiments in this elaborate simulation environment. Experimental results show that our approach reduces the amount of communication up to 71.9% in synthetic benchmarks, and 77% in real-world benchmarks compared to an improved version of our closest rival. Consequently, the overall energy consumption is improved by up to 17.8% in synthetic benchmarks, and 79% in the real-world benchmark, by our technique compared to that improved rivals.

4. OUR APPROACH

In this section, we first show the flow of our work. Since pure VM placement problem is similar to the bin-packing problem in computer science, we use one of corresponding algorithms, the first-fit algorithm, as a baseline algorithm to compare to.

4.1 Our Optimization and Evaluation Flow

Figure 7 gives an overview of our optimization and evaluation flow. First, we use a VM placement algorithm to place the VMs on servers (the *placement phase* in the Fig. 7). We present two algorithms in this paper (see Section 5) corresponding to this step. Then, the results of the placement are evaluated to report here and compare to related work (the *evaluation phase* in the Fig. 7). There are many real-world cases even at large scales where the VMs are basically static and the goal is to optimally place them for the lowest energy. The flow in Fig. 7 is sufficient to address the needs of such cases, but for other cases where the nature of VMs is semi-static or dynamic, we rerun our VM placement algorithm (see Fig. 8) at certain intervals if new VMs arrive or existing VMs finish execution during that interval. We assume 30 min for this reexecution interval in this work. While the choice of this time period for replacement interval may not be optimal, it is short enough to address most dynamic behaviors, while it is long enough to cover rather long boot up/shut down latency of several new/finished VMs, and furthermore, it is long enough to reduce the energy and time overhead of rerunning the placement algorithm—see Section 6.3. Deciding the optimal interval for VM replacement is an interesting objective and is part of our future work.

After the placement phase is accomplished, it is time to evaluate the proposed VM placement so we go on to the evaluation phase in Fig. 7. Since the placement algorithm is re-run at fixed intervals, we evaluate its advantages in a single interval in our experiments in Section 6. The overhead of live

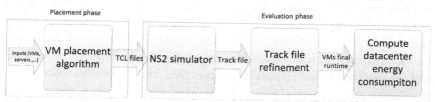

Placement phase   Evaluation phase

Inputs (VMs, servers,...) → VM placement algorithm → TCL files → NS2 simulator → Track file → Track file refinement → VMs final runtime → Compute datacenter energy consumpiton

Figure 7 Our optimization and evaluation overall flow.

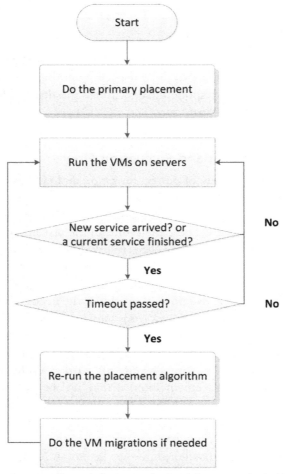

Figure 8 Placement and replacement phases of our technique for dynamic environments.

migration of a number of VMs, if the replacement decides so at the begin-ning of a new interval, is analyzed in Section 6.3. Based on the VM place-ment determined by the algorithm, we generate TCL script files that are the input of the NS2 simulator for network traffic assessment. TCL files consist of information such as network topology and amount of data that transfers between the VMs. After NS2 has finished simulating the network operation, it produces track files. In a track file, the sender, the receiver, the send time, and the receive time of all packets are specified and we use this information to calculate the communication overhead of each VM and add it to the

VM total execution time. Finally, we compute energy consumption of datacenter IT equipment based on the VMs updated execution time.

5. PROBLEM FORMULATION AND ALGORITHMS

In this section, we give notations and formulate the optimization problem. After that, the proposed placement algorithms are described.

5.1 Energy Model

Servers consume a lot of energy in datacenters. Server power consumption is proportional to CPU utilization. Some researches indicate that an idle server consumes around two-thirds of its peak load consumption. The remaining one-third changes almost linearly with the increase in the level of CPU load. According to this, we use the following server power model [69,19,70]:

$$P_{server} = P_{idle} + (P_{max} - P_{idle})*U \tag{2}$$

where P_{max}, P_{idle}, and U are server power consumption at maximum utilization, server power consumption when idle, and server CPU utilization, respectively. It is noteworthy that since the assumption in our benchmarks is that when a CPU is assigned to a VM, that single CPU is fully utilized by that VM, the U parameter for each server is defined here as the number of active cores of the server divided by the total number of cores deployed in the server.

Switches energy consumption is also significant specially when there is a lot of communication between servers. Energy consumption of some powerful switches is more than regular servers, and hence, it is important to consider energy consumption of them in evaluation. Here, we use the Eq. (1) for switches power consumption.

5.2 Notations

We assume that the resource demand for each VM is determined and fixed and does not change during the run time of the VM. For example, a VM needs two cores and 3 GB RAM and these values are constant during the runtime of VM and do not change. We also assume that the communication matrix of VMs is given. Each element of this matrix shows the amount of communication between two VM. This information can be obtained by profiling the VMs during a test period prior to optimization. The above

Table 1 Notation Used in Our Problem Formulation

Parameter	Meaning/Description
M	Number of servers
N	Number of VMs
S	Number of switches
$VMCore_i$	Required number of cores for VM i
$VMRAM_i$	Required amount of memory for VM i
$VMComTime_i$	Communication time of VM i
$ServerCore_i$	Available number of cores in Sever i
$ServerRam_i$	Available amount of RAM in Server i
$VMtoVMcom_{i,j}$	Amount of communication between VM i and VM j in KB
$StoScom_{i,j}$	Amount of communication between Server i and Server j in KB
$VMonServer_{i,j}$	Binary variable indicating whether VM i is placed on Server j or not ($1 =$ is placed, $0 =$ is not placed)
$isServerUp_i$	Binary variable indicating whether Server i is up or not ($1 =$ server is up, $0 =$ server is not up)
$ServerEC_i$	Energy consumption of Server i
$SwitchEC_i$	Energy consumption of Switch i

specifications of all VMs are given parameters to the algorithm. We define the notation listed in Table 1.

5.3 Problem Formulation

Using the notations given in Table 1, the optimization problem is defined as follows:

For a given set of datacenter structural parameters (i.e., M, S, ServerCore vector, ServerRam vector, and the structure of the datacenter network) and a given set of VM parameters (i.e., N, VMCore vector, VMRAM vector, and VMtoVMcomm matrix), minimize total datacenter energy consumption, covering servers and switches, by assigning each VM to one and only one server while considering the communication delay among the VMs as well as capacity constraints on server resources.

Formal equations are listed below:

Equation (3) designates the objective function which is total energy consumption of all servers and switches. Note that in the placement phase (see Fig. 7), we do not compute this equation; our algorithms take the servers and racks structure into account when placing the VMs to minimize the energy, but the amount of this energy is computed in the evaluation phase, where we run NS2 simulations for obtaining the communication time of each VM to finally compute the energy consumption.

Our goal is to reduce the amount of energy that servers and switches consume. But we should pay attention to some constraints while we are trying to achieve this goal. The first and second constraints that are mentioned in Eqs. (4) and (5) deal with constraints on resources of servers. Equation (4) is about number of cores that is available in a server and states that required number of cores of VMs that are placed on each server must be less than or equal to available cores of server. Equation (5) is similar to (4) but instead of cores count, it focuses on memory volume. Each VM must be placed on one and only one server and Eq. (6) shows this constraint.

After placing VMs on servers, the communication volume between servers is calculated by (7) and then by using NS2 simulator, the communication time of each VM is calculated in (8). Note that the NS2 simulation written as $NS2(StoScomm_{k,j})$ in (8) is not a mathematical model since it involves running the simulator and processing its outputs. It merely clarifies the process taken to obtain communication time after placement. We also count the number of up servers by (9). Reducing energy consumption of servers and switches has a direct relationship with results of (7) and (9). If a placement algorithm can reduce the number of up servers and also communication between servers, then the energy consumption of datacenter will be reduced consequently.

Finally, the energy consumption of each server and switch is calculated by (10) and (11). As explained above, these two equations are not part of the *placement phase* but belong to the *evaluation phase*. In these equations, we use the power models that are introduced in (1) and (2), as well as communication time vector *VMComTime* which is calculated by NS2 simulator *after* the placement phase. *VMComTime* indicates the time of the last packet that is sent or received by a VM. At the end of NS2 simulation phase, it reports the sender, receiver, and the time for each packet. Using this information, we calculate the *VMComTime* for each VM as

$$VMComTime = \max(\text{send or receive time of all packets that this}$$
$$\text{VM has sent or received}).$$

CalculateServersEnergy function in (10) works as follows: for each server in each interval, the function checks to see how many of the VMs placed on the server are running in the current interval based on their total running time (communication time + execution time). After obtaining the number of running VMs, the function calculates the utilization of server (U parameter) based on the number of cores that are active by these VMs. Finally, it calculates the energy consumption of the server in the interval by Eq. (10). It is noteworthy that the server is turned off when all the VMs on it are finished. The sum of the energy consumption in each interval gives the total energy consumption of server.

In *CalculateSwitchesEnergy function* in Eq. (11), in order to determine which top of rack switches are ON in an interval, the function checks the servers that are connected to each of the switches (by using the switch id of servers). Note that if none of the servers are ON (meaning that there is no running VM on the servers) then the switch is turned off. However, if one or more servers are turned on, the switch is also turned on and its energy consumption contributes to total energy consumption of the switches. For layer 2 as well as core switches, the process is the same as above except that instead of servers, now the lower layer switches are taken into account to determine on and off switches. Note also that the datacenter network topology is considered in this step. It is the topology that determines which servers are connected to which top of rack switches as well as the connection between different layers of switches. In fact, we use (10) and (11) to evaluate our approach and compare it with the previous ones.

$$Min \sum_{i=1}^{m} ServerEC_i + \sum_{i=1}^{s} SwitchEC_i \tag{3}$$

$$\sum_{j=1}^{n} VMCore_j * VMonServer_{j,i} \leq ServerCore_i, \quad i = 1 \ldots m \tag{4}$$

$$\sum_{j=1}^{n} VMRam_j * VMonServer_{j,i} \leq ServerRam_i, \quad i = 1 \ldots m \tag{5}$$

$$\sum_{j=1}^{m} VMonServer_{i,j} = 1 \tag{6}$$

$$StoScom_{i,j} = \sum_{k=1}^{n} \sum_{l=1}^{n} VMonServer_{k,i} * VMonServer_{l,j} * VMtoVMcom_{k,l}, \tag{7}$$

$$i, j = 1 \ldots m$$

$$VMComTime_i = NS2\left(StoScom_{k,j}\right), \quad i = 1...n, \quad k,j = 1...m \qquad (8)$$

$$isServerUp_i = \begin{cases} 1, \displaystyle\sum_{j=1}^{n} VMonServer_{j,i} > 0 \\[2mm] 0, \displaystyle\sum_{j=1}^{n} VMonServer_{j,i} = 0 \end{cases} \quad i = 1...m \qquad (9)$$

$$ServerEC_i = isServerUp_i * CalculateServersEnergy(VMComTime, P_{server}),$$
$$i = 1...m$$

$$(10)$$

$$SwitchEC_i = CalculateSwtichesEnergy(VMComTime, P_{switch}),$$
$$i = 1...s \qquad (11)$$

5.4 Proposed VM Placement Algorithms

In this section, we describe our proposed VM placement algorithms. The above-explained VM placement problem is NP-hard, and consequently, we propose a meta-heuristic and a heuristic algorithm to solve it. First, we introduce our algorithm that is based on the simulated annealing technique. Then, we describe a heuristic algorithm we have designed for the same problem.

5.4.1 Simulated Annealing-Based VM Placement

Our simulated annealing-based VM placement (SABVMP) technique takes advantage of the first-fit algorithm in its loop. First-fit consolidates VMs in this way: it puts VMs and servers in two queues and starts from the first VM in the VMs queue. Then checks every server from the head of servers queue to see if the server has enough resources and the VM can be placed on that server. If all the VMs are placed, it terminates successfully, but if there is no suitable server for a VM, it aborts. SABVMP iteratively changes the order of VMs in the queue (line 7, Table 2), and then uses first-fit to place the VMs from the queue onto the servers (line 8, Table 2), and then calculates total output traffic of servers (line 9). It also calculates the amount of traffic between each pair of racks, and multiplies this value with a constant representing the cost (time) of communication between those two racks, and then adds this amount to the calculated traffic of servers. The reason behind this formula (lines 4 and 9) is that our experiments show that communication volume among racks has a strong effect on VMs communication time.

Table 2 Pseudo Code of our SABVMP Algorithm

Algorithm 1—VM Placement with SABVMP

1. Set the highest temperature to **Th**, and the coolest temperature to **Tl**
2. Initialize VMs order in queue, **X**
3. Do VM placement using first fit
4. Calculate total output communication of servers (servers traffic + C ⋆ racks traffic), **O**
5. **T = Th**
6. **While** temperature is higher than **Tl**
7.   Randomly change the order of VMs in queue, **X'**
8.   Do VM placement using first fit algorithm
9.   Calculate total output communication of servers (servers traffic + C ⋆ racks traffic), **O'**
10.   **if** (Accept (**O'**, **O**, temperature)) **then X = X'** and **O = O'**
11.   Decrease the temperature **T** by multiplying it by a constant smaller than one
12. **End**

Note the effect of the topology of the datacenter network in this formula; the constant coefficient mentioned above corresponds to the above topology, or in other words, the structure of the available connections among racks. In the rest of the algorithm (line 10), if the above-mentioned amount of total traffic among servers is accepted over previous iterations (line 10), algorithm saves this value as the new best value of simulated annealing and chooses the current order of VMs in queue as the best one.

The function *Accept* in line 10 of algorithm compares the amount of VM communication (among servers in one rack as well as among different racks) in the new order of VMs (O') against the old one (O) and determines whether to accept O' as the new order or not. If O' reduces the above metric, it is always accepted as the new order; but even if it does not, it may still be accepted by a probability based on *temperature* so as to avoid being stuck in *local* minima. In other words, complying with the simulated annealing philosophy of operation, although O' is actually worse than O, this temporary upward move is accepted hoping that future moves find a better minima. This probability of acceptance reduces with *temperature* so that at the beginning, larger parts of the design space are initially explored, but the moves are mostly downward at later stages near the end of the algorithm. By each iteration, the temperature is reduced (line 11) and so does the above probability. The algorithm continues until the temperature reaches its lowest value, *Tl*. The VMs will be placed on servers according to the final order of VMs in the queue. Table 2 shows the SABVMP pseudo code.

5.4.2 Communication-Aware VM Placement

In terms of processing capacity of PMs, the pure VM placement problem is similar to the bin-packing problem in computational complexity theory where objects of different volumes must be packed into a finite number of bins of a certain capacity in a way that minimizes the number of bins used. However, our above communication-aware VM placement (CAVMP) problem has other characteristics, most importantly existence of communication among VMs, which necessitates designing new algorithms. Nevertheless, algorithms developed for the basic bin-packing problem can still be used in part. Our heuristic algorithm also partially uses the first-fit technique, which is among bin-packing algorithms. In this heuristic algorithm, we try to find VMs that are in the same group and then put them on servers. One of the differences between CAVMP and SABVMP is that unlike SABVMP, CAVMP does VM placement only in one pass, and hence, is much faster than SABVMP and can be even used online. Table 3 contains the pseudo code of our CAVMP algorithm. CAVMP works in this way: it first chooses a VM randomly from the set of VMs (called X in Table 3) and inserts it in the queue X' (line 2) and assigns a *Group ID* to it (line 3). After

Table 3 Pseudo Code of CAVMP Algorithm

Algorithm 2—VM Placement with CAVMP

X: initial set of VMs in arbitrary order
X': queue that contains final order of VMs
Group ID: indicates group ID of each VM (It is initialized to zero)
1. Select a VM randomly from **X** and remove it
2. Insert this VM in **X'**
3. Assign Group ID to this VM
4. **While** X is not empty
5.   Find the VM in **X** that has most communication with VMs in **X'**
6.   **If** there is no VM in **X** that has communication with current VMs in **X'**
7.   Choose a VM randomly from **X**
8.   Increase **Group ID**
9.   **End**
10.   Assign Group ID to new VM
11.   Append this new VM to the tail of **X'**
12.   Remove the selected VM from **X**
13. **End**
14. Try to place all the VMs of a group on the same server
15. Try to place the remaining VMs from previous step in the same rack
16. Finally use topology-aware first-fit to place remaining VMs in **X'** on servers

that, it searches for a VM among VMs in X that has the most communication with VMs that are in the X' queue. This new VM then is appended to the tail of the X' queue and the same Group ID as the previous one is assigned to it. If there is no VM in X that has communication with current VMs in X', but X still is not empty, it indicates that all the VMs in a group have already been detected and the remaining VMs do not belong to this group. Thus, the Group ID is increased and again a VM is randomly relocated from X to X'. This process repeats until every VM in X is transferred to the X' queue. At this state, all the VMs that are in the same group have been detected and have the same Group ID (lines 4–13). Now it is time to place the VMs on servers. The placement phase in the CAVMP algorithm has three steps. At the first step, the algorithm tries to place all the VMs of a group on a single server, so it checks the resource demand of each group against available resources of each server to find a suitable match (line 14). After finishing step one, it is expected that some groups are not placed because their resource demand is more than available resources of all individual servers. Now it is time for step two. In this step, CAVMP tries to place the VMs of a group on a single rack. It also tries to place the VMs of a group that have more communication with each other on the same server (line 15). The goal of this step is first, to avoid communication between racks and second, to reduce the communication inside a rack. Finally at the third step, first–fit algorithm is used to place the remaining VMs in X' queue onto servers (line 16). This step is used when the resource demand of a group is more than available resource of all individual racks. This may happen when a group is very large and need a lot of resources, or when the previous groups of VMs have consumed some part of availed resources of racks and conse-quently, no rack has enough resources left. At this step, CAVMP uses a topology-aware variation of the first–fit algorithm to place the VMs such that the VMs with more communication among them are put in the same server, then the same rack, and then the topologically neighbor racks (this is accom-plished by selecting the VMs from the head of the above queue in which the VMs with more communication are put next to each other). Note that this hierarchical placement approach is also the other place where the topology of the datacenter network is taken into account; VMs are assigned to servers and racks considering their topological neighborhood.

6. EXPERIMENTAL RESULTS

In this section, we first describe the topology and specification of the datacenter IT equipment, and then the benchmarks are described and finally the results of our algorithms are compared to the competitors. We compare results of our two algorithms, SABVMP and CAVMP, to three rivals: a baseline communication-unaware first-fit algorithm as well as an improved version of the structural constraint-aware virtual machine placement (SCAVP) technique, which we call SCAVP+, where integer linear programming (ILP) is used to find the best placement of VMs within racks based on the same criteria that SCAVP uses. Note that SCAVP [66] is the most relevant CAVMP work we are aware of. Another rival algorithm with which we compare our algorithms, is the VMFlow approach [65], which is a communication-aware algorithm as well, and is introduced in the related work section.

6.1 Datacenter Network Topology

Three-tier tree topology is a common architecture in today datacenters [71], so we used this architecture for evaluation (see Fig. 9); core switches are at the topmost layer of datacenter network. Next level consists of aggregate switches. Finally at the lowest level, the top-of-rack switches have been placed and servers are connected to them. We assumed 192 servers for the datacenter where every 24 servers are packed in one rack, so we have eight racks in total, and hence, eight top-of-rack switches. There are also four aggregate switches and two core switches in the datacenter as shown in Fig. 9. Each top-of-rack switch is connected to two aggregate switches and each aggregate switch is connected to both core switches. One Gigabit Ethernet (1 GE) links are used for connecting servers to top-of-rack switches. All other links between switches are 10 GE.

In simulations on synthetic benchmarks, we used servers with eight cores and 12 GB RAM, and obtained the values for power consumption of servers from Ref. [72], and for switches from Ref. [25]. All other values we used for

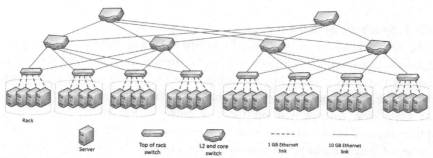

Figure 9 Three-tier tree topology used in most datacenters and our experiments.

Table 4 Values Used for Parameters to Calculating Energy Consumption

Parameter	Value (Watt)
P_{idle}	128
P_{max}	247
$P_{chassis}$ (top of rack switch)	90
$P_{linecard}$ (top of rack switch)	30
P_r (top of rack switch)	1
$P_{chassis}$ (layer 2 and core switches)	520
$P_{linecard}$ (layer 2 and core switches)	35
P_r (layer 2 and core switches)	2

calculating total energy consumption are listed in Table 4. In the real-world benchmark, the VM specifications are based on the available online data and are given separately in Section 6.2.4. Since top-of-rack switches are different from layer 2 and core switches, power values for them are given in different rows in the Table 4. Each switch in our experiments, both in synthetic as well as real-world benchmarks, has one linecard and each linecard has 24 ports, thus we have $n_{linecard} = 1$ and $n_{ports} = 24$.

6.2 Results

We used two benchmark classes in our work: a set of synthetic benchmarks, as well as two real-world benchmarks. We also used NS2 version *ns-allinone-2.35* in our experiments and Ubuntu 10.04 was the operating system of machine that we used for simulations. The results are presented and discussed below.

6.2.1 Synthetic Benchmarks

The first class of experiments is on *synthetic benchmark* where we generated random values for the number of groups, the amount of communication among the VMs in each group, and also for the resource demands of the VMs. For producing these random values, the random generator function of C++ was used which produces *uniformly distributed* random values. Different seed values were used for each benchmark instance of a certain category of benchmarks when producing the random values. Details of these benchmarks are given in Table 5.

Table 5 Specifications of Synthetic Base Point Benchmarks Used in the Experiments

Number of Instances	Pure Runtime (min.)	Number of VMs	Number of Groups	Number of VMs in Each Group	Communication Between VMs (KB)	Number of Cores for Each VM	Amount of RAM for Each VM (GB)
					Random Values		
10	20	100	40	1–4	50,000–100,000	1–4	1–3

The *Pure Runtime* field in Table 5 is pure computation time of each VM and does not include its communication time which experiments show to be roughly 10 min so as to be close to the 30-min period assumed for re-executing the placement algorithm—see Section 4.1. We calculate the communication time for each VM by the NS2 simulator and then add it to the above pure runtime to obtain the actual runtime of each VM. Note that the actual communication time depends on the communication pattern as well as placement of VMs, and hence, due to the randomness of the benchmarks, it is not possible, nor needed, to more accurately tune them to absolutely total 30 min of execution time. It is also noteworthy that in all experiments, a homogeneous datacenter is assumed so the pure runtime of VMs does not change on different servers.

We assume that the VMs are provided in groups and that the VMs in a group only communicate to one another and have no communication with the VMs in other groups. This scenario corresponds to most multi-tier implementations of internet-scale services, where machines form multiple groups or tiers each of which serves a specific part needed for accomplishment of the overall task. Furthermore, in many dynamic environments such as infrastructure-as-a-service (IaaS) cases, independent services run separately and do not communicate with one another.

Total energy consumptions of the five algorithms are given in Table 6 for each benchmark as well as averaged over all of them. As the Table 6 shows, our CAVMP algorithm improves total energy consumption from 15.1% up to 21.0% (16.6% on average) compared to the first-fit algorithm, and from 5.8% up to 13.0% (9.2% on average) compared to the SCAVP+ algorithm. Compared with VMFlow, the CAVMP improves the energy consumption from 5.9% up to 15.1% (10.8% on average). One can see from these results that SCAVP+ is more successful than VMFlow in improving energy consumption. The cause of better improvements in benchmarks 4, 8, and 9 is that in these benchmarks, the amount of communication between VMs is

Table 6 Total Energy Consumption (kWh) Comparison of the Algorithms on Synthetic Benchmarks

| | Algorithm | Benchmarks | | | | | | | | | | Average |
		1	2	3	4	5	6	7	8	9	10	
Total energy consumption (kWh)	First-fit	5.172	5.077	5.107	5.229	5.143	5.11	5.296	5.23	4.963	5.409	5.174
	SCAVP+	4.632	4.518	4.609	4.975	4.627	4.523	4.986	4.695	4.842	5.101	4.751
	VMFlow	4.950	4.782	4.909	4.942	4.831	4.443	4.948	4.866	4.575	5.104	4.835
	SABVMP	4.505	4.312	4.398	4.457	4.444	4.336	4.628	4.388	4.557	4.834	4.486
	CAVMP	4.338	4.118	4.316	4.434	4.361	4.183	4.484	4.131	4.212	4.552	4.313
Improvement (%)	CAVMP versus first-fit	16.1	18.9	15.5	15.2	15.2	18.1	15.3	21	15.1	15.9	16.6
	CAVMP versus SCAVP+	6.4	8.9	6.3	10.9	5.8	7.5	10.1	12	13	10.8	9.2
	CAVMP versus VMFlow	12.4	13.9	12.1	10.3	9.7	5.9	9.4	15.1	7.9	10.8	10.8

more than the other benchmarks. As these numbers are randomly generated, in the mentioned benchmarks these numbers are greater than others and consequently our algorithms have achieved more improvements in energy consumption.

More details of the experimental results on these benchmarks are given in Tables 7 and 8. In Table 7, breakdown of total power consumption into that of servers and switches is provided. The results indicate that CAVMP improves the energy consumption of servers from 10.3% up to 22.3% (16.8% on average) and energy consumption of switches from 15.8% up to 19.6% (17.7% on average) compared to first-fit algorithm. Improvement of CAVMP compared to SCAVP+ for energy consumption of severs is from 8% up to 15.9% (10.4% on average) and for energy consumption of switches is from 0.1% up to 15.7% (5.9% on average).

Table 8 gives network activity details of the benchmarks after being placed on physical servers by the algorithms. The results show that CAVMP reduces number of created packets from 46.2% up to 71.7% (55.4% on average) compared to first-fit algorithm, and from 44.5% up to 71.9% (54.3% on average) compared to SCAVP+ algorithm. For average communication time, CAVMP improves it from 54.5% up to 80.1% (62.7% on average) compared to first-fit, and from 42.6% up to 75% (55.1% on average) compared to SCAVP+ algorithm. The results clearly prove our claims earlier in the paper that too simplistic models for communication overhead are inaccurate and mislead prior placement algorithms.

In Section 3.3.6, we claimed that paying attention only to communication volume is not enough and this factor alone cannot properly show the effectiveness of a VM placement approach. Now here, Table 8 demonstrates validity of this claim. As an example consider the results of benchmark 4 in Table 8. In this benchmark, SABVMP algorithm produces fewer packets than CAVMP (8,306,332 vs. 8,852,241 packets, respectively) while its average communication time is more than CAVMP algorithm (6.76 s vs. 5.63 s, respectively).

6.2.2 Analysis of Sensitivity to the Number of Groups
Each group of VMs represents a service that is providing a service to the users. Assuming that the services work independently, no communication happens among different groups whereas the VMs in each group may communicate to one another to provide the designated service. The number of these groups, or services, can affect the outcome of the placement algorithms. Thus, we designed two more sets of experiments to evaluate this

Table 7 Breakdown of Energy Consumption of Servers and Switches for the Algorithms

		Energy Consumption of Servers (kWh)					Energy Consumption of Switches (kWh)					Total Energy Consumption (kWh)				
		First-Fit	SCAVP+	VMFlow	SABVMP	CAVMP	First-Fit	SCAVP+	VMFlow	SABVMP	CAVMP	First-Fit	SCAVP+	VMFlow	SABVMP	CAVMP
Benchmarks	1	3.589	3.299	3.295	3.173	3.007	1.583	1.333	1.655	1.332	1.331	5.172	4.632	4.950	4.505	4.338
	2	3.454	3.185	3.207	2.983	2.789	1.623	1.333	1.576	1.329	1.329	5.077	4.518	4.782	4.312	4.118
	3	3.485	3.275	3.252	3.066	2.985	1.622	1.334	1.657	1.332	1.331	5.107	4.609	4.909	4.398	4.316
	4	3.605	3.435	3.326	3.125	3.102	1.624	1.54	1.616	1.332	1.332	5.229	4.975	4.942	4.457	4.434
	5	3.561	3.294	3.255	3.113	3.029	1.582	1.333	1.576	1.331	1.332	5.143	4.627	4.831	4.444	4.361
	6	3.487	3.19	3.079	3.005	2.853	1.623	1.333	1.364	1.331	1.33	5.11	4.523	4.443	4.336	4.183
	7	3.672	3.446	3.372	3.295	3.152	1.624	1.54	1.576	1.333	1.332	5.296	4.986	4.948	4.628	4.484
	8	3.607	3.361	3.248	3.056	2.826	1.623	1.334	1.618	1.332	1.305	5.23	4.695	4.866	4.388	4.131
	9	3.341	3.264	3.085	3.102	2.882	1.622	1.578	1.490	1.455	1.33	4.963	4.842	4.575	4.557	4.212
	10	3.785	3.563	3.527	3.418	3.219	1.624	1.538	1.577	1.416	1.333	5.409	5.101	5.104	4.834	4.552

Table 8 The Number of Packets and Average Communication Time in Synthetic Benchmarks

		Number of Created Packets				Average Communication Time (min)				
		First-Fit	SCAVP+	VMFlow	SABVMP	CAVMP	First-Fit	SCAVP+	VMFlow	SABVMP	CAVMP
Benchmarks	1	16,660,024	16,334,225	12,968,821	9,389,696	7,315,590	12.98	10.63	10.47	8.11	4.92
	2	16,202,921	16,340,557	10,994,352	8,249,222	4,998,043	12.82	10.73	9.07	6.71	3.35
	3	16,372,259	15,868,397	10,511,741	10,153,529	7,715,610	12.77	10.82	9.32	8.59	5.57
	4	16,702,210	16,344,139	12,493,124	8,306,332	8,852,241	13.15	10.78	10.1	6.76	5.63
	5	15,979,576	15,475,093	10,798,503	8,224,744	6,830,880	12.95	10.62	8.95	7.35	4.63
	6	17,237,219	16,576,796	11,677,803	8,079,757	8,863,314	13.18	10.45	9.28	7.27	6
	7	16,343,172	16,008,757	12,834,956	9,653,977	7,560,433	12.88	10.67	9.64	7.87	5.09
	8	14,640,857	14,773,285	10,408,409	6,521,748	4,146,000	13.59	10.82	9.25	5.69	2.71
	9	15,688,719	15,191,705	11,438,370	7,457,326	7,630,100	12.89	11.28	9.25	7.36	4.73
	10	15,935,169	15,447,530	12,085,616	9,145,935	8,566,678	12.78	11.03	9.9	8.01	5.73

Algorithms

effect. Similar to the base point case in previous section, each set of experiments consists of 10 benchmarks with random values as in Table 5 produced with different seed values; the two sets of experiments contain 30 and 50 groups, respectively.

Comparing the energy saving results of the cases for 30, 40, and 50 groups in Fig. 10, it is clearly seen that our technique performs more effectively when the number of groups increases for the same number of total VMs. This shows the significance of considering inter-VM communication even when placing VMs among servers inside a single rack; our technique considers this issue whereas SCAVP (and hence SCAVP+) suffice to only inter-rack communication, and hence, our algorithm better packs groups on individual servers to reduce inter-server traffic.

6.2.3 Analysis of Sensitivity to the Amount of Communication

The volume of communication among the VMs in each group is another important factor that can affect how much energy can be saved by the placement techniques. We conducted another two sets of experiments, again with 10 randomly generated benchmarks as before, but with two other ranges of communication volume among their VMs: one set with communication in the range of 5000–10,000 KB of communication between each VM pair, and another one with 100,000–200,000 KB communication size. The summary of results is shown in Fig. 11. As expected, when there is little communication among servers, no more energy can be saved compared to SCAVP+ algorithm since the differentiating factor of our technique is not

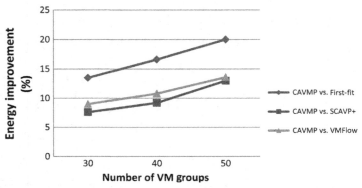

Figure 10 The energy saving of our algorithm improves with the number of VM groups (i.e., the number of services).

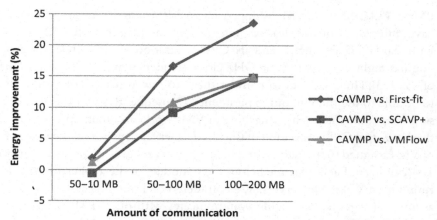

Figure 11 The energy saving of our algorithm improves with the amount of communication among VMs.

seen in the benchmark. However, our technique shines with higher communication volume among VMs.

Another interesting observation in this set of experiments is that our saving is marginally, 0.5%, below that of SCAVP+ at 5–10 MB communication size range. This is because we applied ILP technique in SCAVP+ to improve SCAVP for the purpose of best fitting VMs among servers in a rack, and hence, it outperforms our own heuristic CAVMP placement when there is little to no communication among VMs. However, note that our CAVMP takes only 0.015 s to run on an Intel Corei3 machine (OS: Windows 7, 4 GB of RAM, CPU frequency: 2.93 GHz) in this case whereas SCAVP+ takes two orders of magnitude more time, 3.62 s, for the same case. Thus, our technique is a better choice for *online* usage even in this case.

It is noteworthy that in all the experiments we considered, the CPU utilization of VMs, not servers, to be zero while communicating and 100% while computing, but the utilization of servers depends on the number of VMs placed on them and can be any valid value U.

6.2.4 Real-World Benchmark: Wikipedia Servers

The second class of benchmarks we used is two *real-world benchmarks* obtained from Wikipedia servers [73] and other sources [74–78] which we call *RB2* (*real-world benchmark 2*) from now on for ease of reference. The Wikipedia servers are monitored using Ganglia infrastructure which provides the information of each server such as memory and CPU usage and the amount of communication. In this benchmark, we assumed each

PM of Wikipedia is a VM. Wikipedia actual PMs are of various types and have different resources; for example, one of their machine has 12 CPUs with 2.00 GHz frequency and its CPU utilization is 5%, while another one has eight cores with 2.66 GHz clock frequency and CPU utilization of 9% [73].Thus, we had to convert them to our available cores of VMs to be able to use them in our experiments. Consequently, after the conversion we assume that each virtual core of VMs works at full utilization. Note that the actual CPU utilization of the physical server depends on the number of VMs assigned to that server, and hence the server cores are not necessarily 100% utilized. Each machine belongs to a module of the application that is running on Wikipedia servers. We considered each module as a group and assumed that each PM mainly communicates with other PMs inside that group.

For each PM, Ganglia provides only aggregate input and output traffic volume but does not give the amount of communication between every pair of machines. In order to create the communication matrix, that indicates the amount of communication between every two machines in each interval, we used the simple gravity model [79]. In this model for calculating the traffic between two machines, i and j, the following equation is used:

$$C_{i,j} = \frac{\left(\text{out}_i {}^* \text{in}_j\right)}{\left(\sum_k \text{in}_k\right)}$$

where out_i is output traffic of PM i and in_i is input traffic of PM j [65]. It is noteworthy that although this model cannot accurately determine actual communication among machines since in reality some machines only serve independent requests and do not communicate with other nodes, but due to lack of any further details of communication, there is no more accurate alternative and we had to suffice to this model.

The Wikipedia benchmark has 100 VMs and RB2 has 132 VMs, and the pure runtime (not covering communication time) of all the VMs is 600 min in Wikipedia benchmark and 300 min in the RB2 one. The number of cores that each VM needs in these benchmarks is more than our synthetic ones, so we used bigger servers in simulations of these benchmarks. Here each server has 16 cores and 64 GB memory and for the power parameters, we have $P_{\text{idle}} = 283$ W, $P_{\text{max}} = 388$ W [80]. The power parameters for switches are the same as Table 4.

First, we describe the Wikipedia benchmark. This benchmark consists of 10 groups and each of these groups has its own number of VMs. A variety of resource demands are found in these VMs but their most influential difference is the pattern of communication among these VMs. As can be seen in Table 9, some groups have little communication inside themselves despite their big size, but some other small ones such as 4 and 5 have very high communication. This is important because the energy consumption results that are presented later heavily depend on this communication pattern.

The RB2 benchmark has 19 groups with total 132 VMs. In this benchmark, unlike the previous one, we can only see group No. 9 with small size and high communication and it will affect the energy consumption that we will see later.

The first parameter to compare is the number of up servers. In the Wikipedia benchmark, SCAVP+ and SABVMP algorithms used 31 servers while the first-fit and CAVMP used 32 servers. The difference is more in the RB2 benchmark where SCAVP+ uses the least number of servers with 23 servers, and after that is SABVMP with 24 servers. The third rank in terms of fewer numbers of servers is for first-fit with 26 servers and then comes VMFlow with 28 ones. As we expected, here again the CAVMP uses the most servers by turning on 37 ones. The reason that in the RB2 benchmark we see a bigger difference among the number of used servers by various algorithms is the very high diversity of VMs in terms of demand for resources; this leads to different results in different algorithms depending on how the algorithm decides which server to turn on. Note that CAVMP again uses more servers than SCAVP+ since SCAVP+ takes advantage of an efficient bin-packing-based placement.

To assess how much traffic each algorithm transmits, we count the number of transmitted packets that are transferred over the network of the datacenter. Figure 12 shows the number of packets that each algorithm transmits. The size of each packet is 8 KB. As can be seen, CAVMP transmits fewer packets than any other algorithm and significantly improves the number of transmitted packets by 77% compared to SCAVP+ and 53% compared to VMFlow in the Wikipedia benchmark. The reason for this dramatic decrease in the number of transmitted packets is that CAVMP places all the VMs in groups 4, 5, and 8 on the same servers, and since each of these groups have huge communication inside themselves; this communication volume is offloaded from the datacenter network. In the RB2 benchmark, we can see that the difference between CAVMP and the rival algorithms is less than Wikipedia benchmark since groups such as 4, 5, and 8 from Wikipedia do not exist in this benchmark. Here, the CAVMP decreases the communication volume by 49% and 39% compared to

Table 9 Details of the Communication Pattern of our Real-World Benchmarks

Wikimedia Benchmarks			RB2 Benchmark					
Group No.	Number of VMs	Average Amount of Communication Between VMs in Each Interval (KB)	Group No.	Number of VMs	Average Amount of Communication Between VMs in Each Interval (KB)	Group No.	Number of VMs	Average Amount of Communication Between VMs in Each Interval (KB)
1	37	11,000	1	4	500,000	11	7	700,000
2	31	12,500	2	15	4000	12	2	220,000
3	4	58,000	3	2	50,000	13	15	60,000
4	4	5,000,000	4	8	5000	14	6	64,000
5	4	3,200,000	5	7	25,000	15	16	22,000
6	4	130,000	6	3	50,000	16	6	400,000
7	6	10,000	7	4	15,000	17	4	60,000
8	3	950,000	8	4	14,000	18	16	6000
9	3	40,000	9	5	1,000,000	19	4	210,000
10	4	230,000	10	4	200,000			

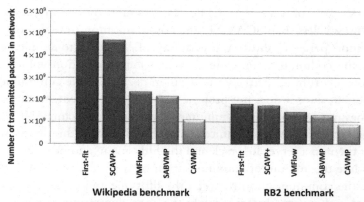

Figure 12 Number of packets each algorithm transmits over the datacenter network in real-world benchmarks.

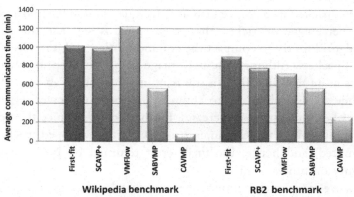

Figure 13 Average VM communication time in each algorithm in our real-world benchmarks.

SCAVP+ and VMFlow, respectively. The number of transmitted packets for SCAVP+ and first-fit are close to each other; however since SCAVP+, unlike first-fit, tries to reduce the communication among the racks, most of its communication is between servers inside the same rack; we will shortly see the effect of it on energy consumption.

Average communication times for VMs in each algorithm are compared in Fig. 13. As the Fig. 13 shows, first-fit performs the most poorly because it is totally communication-unaware. In the Wikipedia benchmark, CAVMP shows the best operation and its average VM communication time is about 93% less than first-fit, about 92% less than SCAVP+, and 93.5% less than

VMFlow. We observe that while the amount of communication in VMFlow was less than SCAVP+, the communication time is more. This observation perfectly shows the disability of VMFlow to simultaneously reduce the communication time as well as communication volume. The VMFlow algorithm is partially successful in decreasing the communication volume, but since it does not propose an intelligent placement, this reduction does not improve the communication time. For example, if a group with a lot of inter-VM communication has five members, this algorithm puts four of them in a server in a rack and the fifth one in another rack. In such case, while the amount of communication between the four VMs is eliminated, the communication between the fifth one and others cause all the switches and servers that are involved in the communication to remain ON and consume energy. We will provide the experimental results of its effects later.

The downside of concentrating all communication in a few servers is the increase in packet drops. Since first-fit distributes communication across the servers, percentage of packets that are dropped is less than three others. Other algorithms try to compact the communication and it results in more dropped packets. Although the percentage of packet drop in other algorithms is more than first-fit, it is still marginal. It is observable in Table 10 that while an algorithm tries to compact the traffic on fewer servers, the amount of dropped packets increases. Further note that since the network protocol used in the experiments is TCP, the dropped packets are retransmitted, and hence, the final result returned by NS2 covers this retransmission time as well.

Now we compare the amount of energy that the placement suggested by each algorithm consumes. As mentioned before, in the experiments we only consider the energy that the IT equipment (servers and switches) consume in the datacenter but do not take into account the energy consumption of other parts such as the cooling system. We calculate the amount of energy that each individual switch and server consumes and according to them calculate the total energy consumption. Results are depicted in Fig. 14 and full details are given in Table 10.

As in synthetic benchmarks, we considered full CPU utilization for VMs during runtime and zero utilization during communication time; utilization of physical servers is based on the number of VMs running on them at each time. For example, if all the VMs on a server are solely communicating in a specific period, the utilization of that server is then considered zero and the server power consumption is only its idle power.

Table 10 Summary of Experimental Results for Real-World Benchmarks

	Wikimedia Benchmark					RB2 Benchmark				
	First-Fit	SCAVP+	VMFlow	SABVMP	CAVMP	First-Fit	SCAVP+	VMFlow	SABVMP	CAVMP
Number of up servers	32	31	31	31	32	26	23	28	24	37
Number of transmitted packets	5.1×10^9	4.7×10^9	2.4×10^9	2.2×10^9	1.1×10^9	1.8×10^9	1.8×10^9	1.5×10^9	1.3×10^9	8.9×10^9
Average VM communication time (min)	1015	987	1223.68	563.36	79.32	901.939	777.705	720.03	565.947	258.697
Packet drop (%)	0.246	0.4263	0.249	0.8163	1.9755	0.027	0.0277	0.0303	0.0383	0.0497
Servers energy consumption (kWh)	504.805	411.061	367.375	239.077	138.99	453.971	362.156	389.164	247.461	164.099
Switches energy consumption (kWh)	445.623	444.099	872.136	374.078	43.8826	635.131	551.286	544.087	487.481	169.17
Total energy consumption (kWh)	950.428	855.16	1239.51	613.156	182.872	1089.1	913.442	933.25	734.942	333.27
Energy reduction compared to SCAVP+ (%)	−11	0	−45	29	79	−19	0	−2	20	64

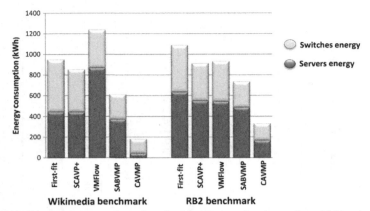

Figure 14 Energy consumption of each algorithm under the real-world benchmarks.

Table 11 Execution Time of Algorithms in Real-World Benchmark in Seconds

	First-Fit	SCAVP	VMFlow	SABVMP	CAVMP
Ganglia	0.088 s	3.58 s	0.481 s	36.069 s	0.444 s

As a result, CAVMP provides the best placement and reduces the energy consumption by 81% compared to first-fit in Wikipedia benchmark and 69% in RB2 benchmark. Comparison of SCAVP + and CAVMP indicates that CAVMP has better results and can save energy consumption by 79% and 63% in Wikipedia benchmark and RB2 benchmark, respectively compared to SCAVP +. This clearly shows the importance of considering inter-VM communication (compare to first-fit), and more importantly, the significance of considering the structure of server as well as racks (compared to SCAVP +), when placing VMs in order to reduce total energy consumption.

Please note that the reason for such a big difference between the improvement of synthetic benchmarks and real-world ones is the huge amount of inter-VM communication in real-world benchmarks. The higher the inter-VM communication, the better the benefits of our approach. Note that if we increase the amount of communication in synthetic benchmarks, the same results as real-world benchmarks will be obtained.

6.2.5 Algorithm Execution Time

Table 11 gives execution time of SABVMP and CAVMP algorithms when run on an Intel Core i3 2.93 GHz machine with 4 GB of memory. Execution time of CAVMP is two orders of magnitude less than SABVMP. Indeed

CAVMP execution time is very low and so it can be effectively used as an online algorithm. Due to absence of any pre-processing at the placement time (such as grouping the VMs that other algorithms do), first–fit algorithm has the least execution time.

6.3 Overhead of Re-Running VM Placement

In the above experiments, we considered a single time interval after placing the VMs, or a semi–static environment with some constant set of services and VMs, such as the real–world Wikipedia servers. However, for more dynamic environments such as IaaS cases, in which new services may start and current services may terminate from time to time, we run the VM placement algorithm periodically as in Fig. 8 and find the new placement based on the new status of VMs. This may impose two overheads that should be assessed here: the overhead of executing the placement algorithm and the overhead of migrating VMs among servers if needed. The execution time of our online algorithm, CAVMP, is negligible (as detailed above and quantified in Table 11), and hence, it has virtually no overhead. The newly run placement algorithm may decide a new place for a number of existing VMs to improve energy consumption in the new interval, and hence, these VMs should be live migrated to other servers. Thus, it is important to evaluate the energy and time overhead of these migrations. The actual number of migrations depends on the actual case and the number and communication pattern of newly arrived VMs as well as that of finished ones. We provide a worst–case analysis here to show the overhead is negligible even if all VMs have to undergo a live migration at the beginning of the new interval.

The required time to do a live migration can be calculated as the VM's size of memory divided by available network bandwidth [81]. This is because the data and images of VMs are usually stored on a network attached storage (NAS) so there is no need to move them among servers [81]. Considering n VMs with M GB of memory each (worst case in Table 5), nM GB should be moved among m servers which provide m Gbps bandwidth assuming 1 Gbps Ethernet links. Thus, it takes $8 \times M \times n/m$ s to migrate all n VMs. In other words, for the values in synthetic benchmarks as in Table 5, it takes around $8 \times 3 \times 100/30 = 80$ s to do all migrations if in the worst case all VMs were decided to be moved. This represents worst case overhead of negligibly 4.4% if the replacement interval is 30 min as assumed in Section 4.1. We repeat that deciding the optimal interval for VM replacement is an interesting objective that we intend to explore as a future direction for further research.

7. CONCLUSION AND FUTURE WORK

Datacenters' voracious appetite for energy has motivated researchers around the globe to find techniques and mechanisms for quenching it. Cooling system and IT equipment are of paramount importance regarding energy consumption in datacenters. IT equipment itself can be split into computing resources and communication resources. The focal point of this chapter is energy efficiency of networking equipment in datacenters. The chapter has surveyed and categorized various approaches in this area and tried to present them in a clear picture. Finally, a new approach by the authors of chapter is illustrated that tries to reduce the energy consumption of datacenters by considering communication among the VMs.

In the following, we indicate some future directions regarding reducing usage of equipment which is the main concern of this chapter. In the VDC placement, one important challenge that needs to be addressed is the resource fragmentation. The current approaches just consider the directly connected nodes instead of the whole topology of physical infrastructure. So, the resource fragment problems are inevitable in current state. Further research in this area and trying to investigate the network topology can lead to better placement and better use of resources.

Regarding hypervisor enhancement, one future direction can be considering multi-core systems instead of single core ones. Simplifying the problem of resource scheduling among guest domains to cases where there is just one physical core cannot satisfy the modern systems with several cores. Combining software solutions such as aggregating flows with hardware advances such as multiqueue devices can further improve the network performance of virtualized systems. So, it can be another direction for future research in the scope of hypervisor enhancement. Finally, making the network I/O virtualization approaches adaptive to heterogeneous workloads can be another avenue for subsequent researches.

New topologies that try to enhance the network performance in datacenters need more attention. Although some routing algorithms are proposed for them, there is still space for proposing more efficient and practical routing algorithms. Various aspects of these newly proposed architectures such as bisection bandwidth or their incremental nature need more investigations. There are also opportunities for designing new architectures based on servers that have more than two NICs.

Considering the heterogeneous network equipment and try to utilize them in an efficient way by traffic engineering and co-locating different flows can be a direction for future researches. Some of the current approaches just consider the state of switches (being ON/OFF) or their active time. However, the bit rate of ports as well as the number of used/unused ports of switch can also affect the decision making in traffic engineering and bring some valuable opportunities for energy saving.

In addition, there are a number of avenues of research we propose to follow as future directions regarding our approach and other communication-aware consolidation approaches. One open challenge in this work is to find the optimal solution of the problem. To reach this goal it is necessary to involve an estimation of network communication time in the placement phase. Results of NS2 simulation cannot be directly used here since the execution time of each NS2 simulation is very high (up to several hours). Finding a way to estimate the communication time without needing NS2 simulations per iteration is necessary here.

Finally, considering and evaluating the effects of other datacenter structural features such as cooling structure and thermal effects (e.g., rack-based vs. row-based vs. room-based cooling), network topologies, power distribution mechanisms (e.g., modular distributed UPS vs. centralized UPS), as well as application behaviors (e.g., burst network usage in cases such as shuffling phase of MapReduce jobs) and their effects and interaction with VM placement are other interesting questions to seek answers for.

ACKNOWLEDGMENTS

This work was supported in part by the Iran Telecommunication Research Center Grant no. 500/12188/t and also the Research Grant Program of Sharif University of Technology.

Seyed Morteza Nabavinejad would like to thank Mr. Pahlavan, Mr. Momeni, and Ms. Falahati who assisted in preparing this chapter.

REFERENCES

[1] Google. http://www.google.com/about/datacenters/efficiency/internal, 2014 (accessed November 10th, 2014).
[2] F. Ahmad, T.N. Vijaykumar, Joint optimization of idle and cooling power in data centers while maintaining response time, in: Proceedings of the Fifteenth Edition of Asplos on Architectural Support for Programming Languages and Operating Systems, 2010, pp. 243–256.
[3] V.K. Arghode, Y. Joshi, Modeling strategies for air flow through perforated tiles in a data center, IEEE Trans. Compon. Packag. Manuf. Technol. 3 (2013) 800–810.
[4] Q. Tang, S.K.S. Gupta, G. Varsamopoulos, Energy-efficient thermal-aware task scheduling for homogeneous high-performance computing data centers: a cyber-physical approach, IEEE Trans. Parallel Distrib. Syst. 19 (2008) 1458–1472.

[5] A. Sansottera, P. Cremonesi, Cooling-aware workload placement with performance constraints, Perform. Eval. 68 (2011) 1232–1246.

[6] A. Pahlavan, M. Momtazpour, M. Goudarzi, Power reduction in HPC data centers: a joint server placement and chassis consolidation approach, J. Supercomput. 70 (2014) 845–879.

[7] Facebook. https://www.facebook.com/notes/facebook-engineering/designing-a-very-efficient-data-center/10150148003778920, 2014 (accessed November 10th, 2014).

[8] D. Wong, M. Annavaram, Scaling the energy proportionality wall with KnightShift, IEEE Micro 33 (2013) 28–37.

[9] L. Tan, C. Minghua, L.L.H. Andrew, Simple and effective dynamic provisioning for power-proportional data centers, IEEE Trans. Parallel Distrib. Syst. 24 (2013) 1161–1171.

[10] L. Minghong, A. Wierman, L.L.H. Andrew, E. Thereska, Dynamic right-sizing for power-proportional data centers, IEEE/ACM Trans. Networking 21 (2013) 1378–1391.

[11] A. Krioukov, P. Mohan, S. Alspaugh, L. Keys, D. Culler, R. Katz, NapSAC: design and implementation of a power-proportional web cluster, SIGCOMM Comp. Comm. Rev 41 (2011) 102–108.

[12] N. Bobroff, A. Kochut, K. Beaty, Dynamic placement of virtual machines for managing SLA violations, in: 10th IFIP/IEEE International Symposium on Integrated Network Management, 2007, pp. 119–128.

[13] L.A. Barroso, U. Holzle, The case for energy-proportional computing, Computer 40 (2007) 33–37.

[14] F. Hermenier, X. Lorca, J.-M. Menaud, G. Muller, J. Lawall, Entropy: a consolidation manager for clusters, in: Proceedings of the ACM SIGPLAN/SIGOPS International Conference on Virtual Execution, Environments, 2009, pp. 41–50.

[15] M. Marzolla, O. Babaoglu, F. Panzieri, Server consolidation in Clouds through gossiping, in: IEEE International Symposium on World of Wireless, Mobile and Multimedia Networks (WoWMoM), 2011, pp. 1–6.

[16] X. Wang, Z. Liu, An energy-aware VMs placement algorithm in Cloud Computing environment, in: International Conference on Intelligent System Design and Engineering Application (ISDEA), 2012, pp. 627–630.

[17] H. Goudarzi, M. Ghasemazar, M. Pedram, SLA-based optimization of power and migration cost in cloud computing, in: IEEE/ACM International Symposium on Cluster, Cloud and Grid Computing (CCGrid), 2012, pp. 172–179.

[18] S. Weiming, H. Bo, Towards profitable virtual machine placement in the data center, in: IEEE International Conference on Utility and Cloud Computing (UCC), 2011, pp. 138–145.

[19] D. Kliazovich, P. Bouvry, S.U. Khan, DENS: data center energy-efficient network-aware scheduling, in: IEEE/ACM Int'l Conference on & Int'l Conference on Cyber, Physical and Social Computing Green Computing and Communications (GreenCom), (CPSCom), 2010, pp. 69–75.

[20] A. Corradi, M. Fanelli, L. Foschini, VM consolidation: a real case based on OpenStack cloud, Futur. Gener. Comput. Syst. 32 (2014) 118–127.

[21] S. Esfandiarpoor, A. Pahlavan, M. Goudarzi, Structure-aware online virtual machine consolidation for datacenter energy improvement in cloud computing, Comput. Electr. Eng. 42 (2015) 74–89.

[22] A. Greenberg, J. Hamilton, D.A. Maltz, P. Patel, The cost of a cloud: research problems in data center networks, SIGCOMM Comp. Comm. Rev. 39 (2008) 68–73.

[23] D. Abts, M.R. Marty, P.M. Wells, P. Klausler, H. Liu, Energy proportional datacenter networks, in: Proceedings of the 37th Annual International Symposium on Computer Architecture (ISCA), 2010, pp. 338–347.

[24] L.A. Barroso, J. Clidaras, U. Hlzle, The Datacenter as a Computer: An Introduction to the Design of Warehouse-Scale Machines, Morgan & Claypool Publishers, San Rafael, CA, 2013.

[25] P. Mahadevan, P. Sharma, S. Banerjee, P. Ranganathan, A power benchmarking framework for network devices, in: Networking, 2009, pp. 795–808.

[26] C. Eddington, InfiniBridge: an InfiniBand channel adapter with integrated switch, IEEE Micro 22 (2002) 48–56.

[27] B. Dickov, M. PericÃ s, P.M. Carpenter, N. Navarro, E. AyguadÃ, Analyzing performance improvements and energy savings in infiniband architecture using network compression, in: Proceedings of the IEEE 26th International Symposium on Computer Architecture and High Performance Computing, 2014, pp. 73–80.

[28] V. Sundriyal, M. Sosonkina, A. Gaenko, Z. Zhang, Energy saving strategies for parallel applications with point-to-point communication phases, J. Parallel Distrib. Comput. 73 (2013) 1157–1169.

[29] C. Decusatis, Optical interconnect networks for datacom and computercom, in: Optical Fiber Communication Conference and Exposition and the National Fiber Optic Engineers Conference (OFC/NFOEC), 2013, pp. 1–33.

[30] D. Larrabeiti, P. Reviriego, J.A. Hern, J.A. Maestro Ndez, M. Urue, Towards an energy efficient 10 GB/s optical ethernet: performance analysis and viability, Opt. Switch. Netw. 8 (2011) 131–138.

[31] R. Danping, L. Hui, J. Yuefeng, Power saving mechanism and performance analysis for 10 gigabit-class passive optical network systems, in: IEEE International Conference on Network Infrastructure and Digital Content, 2010, pp. 920–924.

[32] P. Mahadevan, S. Banerjee, P. Sharma, Energy proportionality of an enterprise network, in: Proceedings of the First ACM SIGCOMM Workshop on Green Networking, 2010, pp. 53–60.

[33] C. Yiu, S. Singh, Merging traffic to save energy in the enterprise, in: Proceedings of the 2nd International Conference on Energy-Efficient Computing and Networking, 2011, pp. 97–105.

[34] A. Carrega, S. Singh, R. Bolla, R. Bruschi, Applying traffic merging to datacenter networks, in: Proceedings of the 3rd International Conference on Future Energy Systems: Where Energy, Computing and Communication Meet, 2012. Article No. 3.

[35] Q. Yi, S. Singh, Minimizing energy consumption of FatTree data center networks, SIGMETRICS Perform. Eval. Rev. 42 (2004) 67–72.

[36] C. Gunaratne, K. Christensen, B. Nordman, S. Suen, Reducing the energy consumption of ethernet with adaptive link rate (ALR), IEEE Trans. Comput. 57 (2008) 448–461.

[37] W. Fisher, M. Suchara, J. Rexford, Greening backbone networks: reducing energy consumption by shutting off cables in bundled links, in: Proceedings of the First ACM SIGCOMM Workshop on Green Networking, 2010, pp. 29–34.

[38] Q. Xiao, J. Hong, A. Manzanares, R. Xiaojun, Y. Shu, Communication-aware load balancing for parallel applications on clusters, IEEE Trans. Comput. 59 (2010) 42–52.

[39] Z. Jidong, S. Tianwei, H. Jiangzhou, C. Wenguang, Z. Weiming, Efficiently acquiring communication traces for large-scale parallel applications, IEEE Trans. Parallel Distrib. Syst. 22 (2011) 1862–1870.

[40] D.M. Divakaran, L. Tho Ngoc, M. Gurusamy, An online integrated resource allocator for guaranteed performance in data centers, IEEE Trans. Parallel Distrib. Syst. 25 (2014) 1382–1392.

[41] S. Luo, H. Yu, L. Li, D. Liao, G. Sun, Traffic-aware VDC embedding in data center: a case study of fattree, China Commun. 11 (2014) 142–152.

[42] W. Xiaohui, L. Hongliang, Y. Kun, Z. Lei, Topology-aware partial virtual cluster mapping algorithm on shared distributed infrastructures, IEEE Trans. Parallel Distrib. Syst. 25 (2014) 2721–2730.

[43] A. Amokrane, M.F. Zhani, R. Langar, R. Boutaba, G. Pujolle, Greenhead: virtual data center embedding across distributed infrastructures, IEEE Trans. Cloud Comput. 1 (2013) 36–49.

[44] M. Bourguiba, K. Haddadou, I. El Korbi, G. Pujolle, Improving network I/O virtualization for cloud computing, IEEE Trans. Parallel Distrib. Syst. 25 (2014) 673–681.

[45] G. Haibing, M. Ruhui, L. Jian, Workload-aware credit scheduler for improving network I/O performance in virtualization environment, IEEE Trans. Cloud Comput. 2 (2014) 130–142.

[46] G. Bei, W. Jingzheng, W. YongJi, S.U. Khan, CIVSched: a communication-aware inter-VM scheduling technique for decreased network latency between co-located VMs, IEEE Trans. Cloud Comput. 2 (2014) 320–332.

[47] S. Govindan, C. Jeonghwan, A.R. Nath, A. Das, B. Urgaonkar, S. Anand, Xen and Co.: communication-aware CPU management in consolidated Xen-based hosting platforms, IEEE Trans. Comput. 58 (2009) 1111–1125.

[48] G. Qu, Z. Fang, J. Zhang, S. Zheng, Switch-centric data center network structures based on hypergraphs and combinatorial block designs, IEEE Trans. Parallel Distrib. Syst. 26 (2015) 1154–1164.

[49] G. Deke, C. Tao, L. Dan, L. Mo, L. Yunhao, C. Guihai, Expandable and cost-effective network structures for data centers using dual-port servers, IEEE Trans. Comput. 62 (2013) 1303–1317.

[50] L. Yong, Y. Dong, G. Lixin, DPillar: scalable dual-port server interconnection for data center networks, in: Proceedings of 19th International Conference on Computer Communications and Networks (ICCCN), 2010, pp. 1–6.

[51] D. Li, J. Wu, On data center network architectures for interconnecting dual-port servers, IEEE Trans. Comput. 64 (2015) 3210–3222.

[52] L.N. Bhuyan, D.P. Agrawal, Generalized hypercube and hyperbus structures for a computer network, IEEE Trans. Comput. C-33 (1984) 323–333.

[53] G. Panchapakesan, A. Sengupta, On a lightwave network topology using Kautz digraphs, IEEE Trans. Comput. 48 (1999) 1131–1137.

[54] F.T. Leighton, Introduction to Parallel Algorithms and Architectures: Array, Trees, Hypercubes, Morgan Kaufmann Publishers Inc., Burlington, MA, 1992

[55] Y. Shang, D. Li, M. Xu, J. Zhu, On the network power effectiveness of data center architectures, IEEE Trans. Comput. 64 (2015) 3237–3248.

[56] C. Guo, H. Wu, K. Tan, L. Shi, Y. Zhang, S. Lu, Dcell: a scalable and fault-tolerant network structure for data centers, in: Proceedings of the ACM SIGCOMM Conference on Data, Communication, 2008, pp. 75–86.

[57] L. Dan, G. Chuanxiong, W. Haitao, K. Tan, Z. Yongguang, L. Songwu, FiConn: using backup port for server interconnection in data centers, in: IEEE INFOCOM, 2009, pp. 2276–2285.

[58] A.Q. Lawey, T.E.H. El-Gorashi, J.M.H. Elmirghani, Distributed energy efficient clouds over core networks, J. Lightwave Technol. 32 (2014) 1261–1281.

[59] W. Lin, Z. Fa, J. Arjona Aroca, A.V. Vasilakos, Z. Kai, H. Chenying, L. Dan, L. Zhiyong, GreenDCN: a general framework for achieving energy efficiency in data center networks, IEEE J. Sel. Areas Commun. 32 (2014) 4–15.

[60] D. Li, Y. Yu, W. He, K. Zheng, B. He, Willow: saving data center network energy for network-limited flows, IEEE Trans. Parallel Distrib. Syst. 26 (2015) 2610–2620.

[61] M. Al-Fares, S. Radhakrishnan, B. Raghavan, N. Huang, A. Vahdat, Hedera: dynamic flow scheduling for data center networks, in: Proceedings of the 7th USENIX Conference on Networked Systems Design and Implementation, 2010, p. 19.

[62] T.C. Ferreto, M.A.S. Netto, R.N. Calheiros, C.A.F. De Rose, Server consolidation with migration control for virtualized data centers, Futur. Gener. Comput. Syst. 27 (2011) 1027–1034.

[63] M. Cao Le Thanh, M. Kayashima, Virtual machine placement algorithm for virtualized desktop infrastructure, in: IEEE International Conference on Cloud Computing and Intelligence Systems (CCIS), 2011, pp. 333–337.

[64] X. Liao, H. Jin, H. Liu, Towards a green cluster through dynamic remapping of virtual machines, Futur. Gener. Comput. Syst. 28 (2012) 469–477.

[65] V. Mann, A. Kumar, P. Dutta, S. Kalyanaraman, VMFlow: leveraging VM mobility to reduce network power costs in data centers, in: Proceedings of the International IFIP TC 6 Conference on Networking-Volume, Part I, 2011, pp. 198–211.

[66] D. Jayasinghe, C. Pu, T. Eilam, M. Steinder, I. Whally, E. Snible, Improving performance and availability of services hosted on IaaS clouds with structural constraint-aware virtual machine placement, in: IEEE International Conference on Services Computing (SCC), 2011, pp. 72–79.

[67] W. Meng, M. Xiaoqiao, Z. Li, Consolidating virtual machines with dynamic bandwidth demand in data centers, in: Proceedings IEEE INFOCOM, 2011, pp. 71–75.

[68] M. Xiaoqiao, V. Pappas, Z. Li, Improving the scalability of data center networks with traffic-aware virtual machine placement, in: Proceedings IEEE INFOCOM, 2010, pp. 1–9.

[69] G. Chen, W. He, J. Liu, S. Nath, L. Rigas, L. Xiao, F. Zhao, Energy-aware server provisioning and load dispatching for connection-intensive internet services, in: Proceedings of the 5th USENIX Symposium on Networked Systems Design and Implementation, 2008, pp. 337–350.

[70] S. Rivoire, P. Ranganathan, C. Kozyrakis, A comparison of high-level full-system power models, in: Proceedings of the Conference on Power Aware Computing and Systems, 2008, p. 3.

[71] OL-11565-01, Cisco Data Center Infrastructure 2.5 Design Guides, 2007.

[72] Dell PowerEdge R710 featuring the Dell Energy Smart 870W PSU and Intel Xeon E5620, ENERGY STAR Power and Performance Data Sheet.

[73] Ganglia Monitoring Tool. Available online at http://ganglia.wikimedia.org/, 2014 (accessed July 6th, 2014).

[74] http://neos-server.org/ganglia/, 2014 (accessed July 6th, 2014).

[75] http://ganglia.it.pasteur.fr/, 2014 (accessed July 6th, 2014).

[76] http://www.meteo.unican.es/ganglia/, 2014 (accessed July 6th, 2014).

[77] https://ganglia.surfsara.nl/, 2014 (accessed July 6th, 2014).

[78] https://dev.monitor.orchestra.med.harvard.edu/, 2014 (accessed July 6th, 2014).

[79] Y. Zhang, M. Roughan, N. Duffield, A. Greenberg, Fast accurate computation of large-scale IP traffic matrices from link loads, in: Proceedings of ACM SIGMETRICS Performance Evaluation Review, 2012, pp. 206–217.

[80] IBM Power 710 Express and IBM Power 730 Express (8231-E2B), ENERGY STAR Power and Performance Data Sheet.

[81] A. Beloglazov, J. Abawajy, R. Buyya, Energy-aware resource allocation heuristics for efficient management of data centers for Cloud computing, Futur. Gener. Comput. Syst. 28 (2012) 755–768.

ABOUT THE AUTHORS

Seyed Morteza Nabavinejad is a Ph.D. student at the Department of Computer Engineering, Sharif University of Technology, Tehran, Iran. He received the B.Sc. degree in Computer Engineering from Ferdowsi University of Mashhad and the M.Sc. from Sharif University of Technology in 2011 and 2013, respectively. He is currently working at Energy Aware Systems (EAYS) Laboratory under supervision of Dr. Maziar Goudarzi. His research interests include BigData analytics, cloud computing, green computing, and energy aware datacenters.

Maziar Goudarzi is an Assistant Professor at the Department of Computer Engineering, Sharif University of Technology, Tehran, Iran. He received the B.Sc., M.Sc., and Ph.D. degrees in Computer Engineering from Sharif University of Technology in 1996, 1998, and 2005, respectively. Before joining Sharif University of Technology as a faculty member in September 2009, he was a Research Associate Professor at Kyushu University, Japan from 2006 to 2008, and then a member of research staff at University College Cork, Ireland in 2009. His current research interests include architectures for large-scale computing systems, green computing, hardware–software codesign, and reconfigurable computing. Dr. Goudarzi has won two best paper awards, published several papers in reputable conferences and journals, and served as member of technical program committees of a number of IEEE, ACM, and IFIP conferences including ICCD, ASP-DAC, ISQED, ASQED, EUC, and IEDEC among others.

AUTHOR INDEX

Note: Page numbers followed by "*f*" indicate figures, and "*t*" indicate tables.

SUBJECT INDEX

Note: Page numbers followed by "*f*" indicate figures, "*t*" indicate tables, and "*b*" indicate boxes.

CONTENTS OF VOLUMES IN THIS SERIES

Volume 71

Volume 72

Volume 73

Volume 94

Volume 95

Volume 96

Volume 97

Printed in the United States
By Bookmasters